Patric

Text Mining for Genomics-based Drug Discovery

Patrick Herron

Text Mining for Genomics-based Drug Discovery

VDM Verlag Dr. Müller

Imprint

Bibliographic information by the German National Library: The German National Library lists this publication at the German National Bibliography; detailed bibliographic information is available on the Internet at http://dnb.d-nb.de.

Cover image: www.purestockx.com

Publisher:
VDM Verlag Dr. Müller Aktiengesellschaft & Co. KG
Dudweiler Landstr. 125 a, 66123 Saarbrücken, Germany
Phone +49 681 9100-698, Fax +49 681 9100-988, Email: info@vdm-verlag.de

Produced in USA and UK by:
Lightning Source Inc., La Vergne, Tennessee, USA
Lightning Source UK Ltd., Milton Keynes, UK

ISBN: 978-3-8364-3714-1

1

Contents

Table of Contents

5

Table of Figures

The best way to predict the future is to invent it.

  - Alan Kay

Innovation is not the product of logical thought, although the result is tied to logical structure.

  - Albert Einstein

The library is unlimited but periodic.

  - Jorge Luis Borges, "The Library of Babel"

I. Introduction

Through innovation, drug companies prolong and improve human life. How pharmaceutical companies innovate is the story of how they find new drug treatments. The largest of drug companies however do not produce many drugs every year yet they spend billions trying. When the Human Genome Project brought with it the hope of revolutionizing medicine through massive amounts of new genetic data, pharmaceutical companies worked quickly to help found the field known as *pharmacogenomics*. Drug companies invested in pharmacogenomics hoping to improve their abilities to innovate.

Years after the completion of the Human Genome Project, however, pharmacogenomics-based drug discovery finds itself awash in a sea of data with little of the anticipated dramatic success. While bioinformatics researchers have labored to reduce the information overload in pharmacogenomics-based using text mining, two questions remain. First, exactly how can text mining help pharmaceutical companies discover drugs via pharmacogenomics? Secondly, how are pharmaceutical companies innovating or adopting text mining technologies?

The purpose of the present study is to develop a better understanding of text mining and its role in pharmacogenomics-based drug discovery. To that end, I first evaluate the role of innovation in the pharmaceutical industry as well as the rise of pharmacogenomics-based drug discovery. I then define text mining in a comprehensive way, distinguishing it from information extraction by emphasizing the centrality of generating novel information. Next, I use the new text mining framework to evaluate successful business and scientific text mining applications. I then perform a case study, employing informal interviews of key drug discovery and informatics decision-makers from a large pharmaceutical corporation. Analyzing the case study provides me with a real-world grounding for understanding text mining adoption and innovation, particularly with respect to pharmacogenomics-based drug discovery. I frame the discussion of text mining adoption and innovation using the Unified Theory of Acceptance and Use of Technology (UTAUT) model (Venkatesh, Morris, Davis, & Davis, 2003) as well as Everett Rogers's Diffusion of innovations theory (2003).

Ultimately, I attempt to develop a picture of text mining in the pharmaceutical industry, both where it is and in what direction it may head. I further hope to provide a clearer illustration of text mining itself and its relationship to similar yet distinct types of technologies. More generally, I wish to gain some insight into the relationship between adoption and innovation.

II. Innovation and Drug Discovery in the Pharmaceutical Industry

Pharmaceutical corporations are constantly in need of innovation. The discovery of new drug treatments, particularly new molecular entities (NMEs), drives the pharmaceutical industry (Accenture, 2003). The discovery of new drugs and drug treatments helps pharmaceutical companies work towards the central goal of medicine: improving and prolonging life (Lichtenberg, 1998). As pharmaceutical companies struggle to gain competitive advantage, the importance of innovation increases (Cardinal, 2001).

Market-ready NMEs, arguably the main drivers of both business and social goals of pharmaceutical organizations, are historically difficult for pharmaceuticals to come by. In 2005 only 13 NMEs were approved for medical use by the US Federal Government (US Food and Drug Administration Center for Drug Evaluation and Research, 2005b), and in 2004, 31 were approved (US Food and Drug Administration Center for Drug Evaluation and Research, 2005a). According to the US Food and Drug Administration Center on Drug Evaluation and Research (CDER), the world's two largest pharmaceutical companies, Pfizer and Glaxo Smith Kline, produced only one NME each over that two-year period while spending a combined US$25 billion on research and development (GlaxoSmithKline, 2006; Pfizer, 2006). The year 2003 saw production of approved

NMEs lower than in the two decades preceding it (GlaxoSmithKline, 2005). Neither

Pfizer nor GSK had NMEs among the 21 NMEs approved that year.

Recent scarcity of NMEs reaching market is not a new phenomenon. Only five

pharmaceutical firms during the 1990s innovated 10 or more approved NMEs (National

Institute for Health Care Management Foundation, 2002). Based on evidence from a

research survey of drug development in the 1990s, the cost of development for each drug

reaching the marketplace is fast approaching $1 billion (DiMasi, 1999) though that figure

has been contested (see Scherer, 2004, or Light & Warburton, 2005, for examples).

The pressure on companies to discover new drugs is on the rise. According to a leading

industry report,

> [t]o match investor expectations 'Big Pharma' needs to double the number of
> NMEs entering clinical development, improve clinical success rates from 10:1 to
> 10:3, and reduce the time from first dose in a human to regulatory approval by 33
> percent [....] It seems that the industry is at a point of saturation where increasing
> the amount of money thrown at a project is not increasing the returns in a
> commensurate fashion. (Accenture, 2003)

The widening of the gap between rising R&D costs and shrinking drug-to-market

numbers is so striking that the phenomenon is sometimes referred to by industry

professionals as the "innovation gap" and the "valley of death" (for examples, see BTG,

2006, and Goldman, 2003). The future of the pharmaceutical industry and of medicine as

a whole greatly depends on pharmaceutical companies regularly and efficiently

producing NMEs.

III. Drug Discovery and Pharmacogenomics

A. Pharmaceutical pipeline

Pharmaceutical companies are in the business of discovering, developing, evaluating, and selling new medications. The general process of bringing drugs to market involves four crucial steps:

1. Drug discovery

2. Drug development

3. Clinical trials in humans

4. Drug marketing (Ng, 2004)

The pipeline that brings NMEs and other pharmaceutical compounds to market begins with drug discovery. It is crucial to note that in the present study "discovery" is being considered equivalent to "innovation;" both terms denote the process of creating something both novel and valuable. Drug discovery serves as the nexus of innovation from which all pharmaceutical organizations build. Drug discoveries fuel a pharmaceutical company because the innovation necessary for creating NMEs drives further development and evaluation down the pharmaceutical pipeline. It should not be surprising, then, that the largest pharmaceutical companies dedicate the lion's share of their research and development budgets to the drug discovery step. The present study therefore will focus on the drug discovery phase of the pharmaceutical pipeline

Drug discovery generally proceeds in a four-step process:

1. identification of medical needs (definition of medical problem)

2. identification and validation of drug targets (receptors) integral to disease processes

3. discovery of lead compounds that interact with target(s)

4. optimize lead compounds for generation of patentable NMEs (Ng, 2004; A. D. Roses, Burns, Chissoe, Middleton, & Jean, 2005)

Proper identification of medical needs translates directly into the effective identification of a disease and its underlying etiology. Likewise, clarifying medical needs also equates with enumerating justifications for treating a disease. Identification of drug targets essentially translates into identifying the loci of disease activity such as cell receptors of the genes that give rise to such receptors. Lead compound identification is equivalent to identifying compound(s) that act upon the drug target(s) preferably in such a way as to either cure the disease or ameliorate all or some of its symptoms. The lead compounds are refined into drug candidates at the optimization stage. The refinement process in the optimization stage often involves small modifications to the lead compound's pharmacokinetics (potency, selectivity, efficacy, bioavailability, and/or metabolic stability) and reductions in toxicity (A. D. Roses et al., 2005).

As noted earlier, the drug discovery pipeline is highly inefficient. Approximately 99.98% of lead compounds fail to make it through the pipeline and produce approved NMEs for medical use; over 50% of lead compounds fail during the refinement phase alone

13

(Cunningham, 2000). Given the rate of failure in conjunction with the cost of success, it

should come as little surprise that industry professionals refer to the problem either as the

"innovation gap" or as the "valley of death." See Figure I.1 below for an illustration of

the pipeline (Gad, 2005, p. 2).

Time (In Years)

16	Product Approved for Market	1
15		
14	FDA Review	2
13	NDA Filed	
12		6
11	Phase III Clinical Trials	
10		8
9	Phase II Clinical Trials	
8		250
7	IND Filed/Phase I Clinical Trials	
6	Proposed for Preclinical Development	400
4		
2	New Compounds from Research	5,000-10,000
0	"Discovery"	

Figure I.1 Attrition during the development of new molecules with a promise of therapeutic potential. Over the course of taking a new molecular entity through scale-up, safety, and efficacy testing, and, finally, to market, typically only 1 out of every 9000 to 10,000 will go to the marketplace.

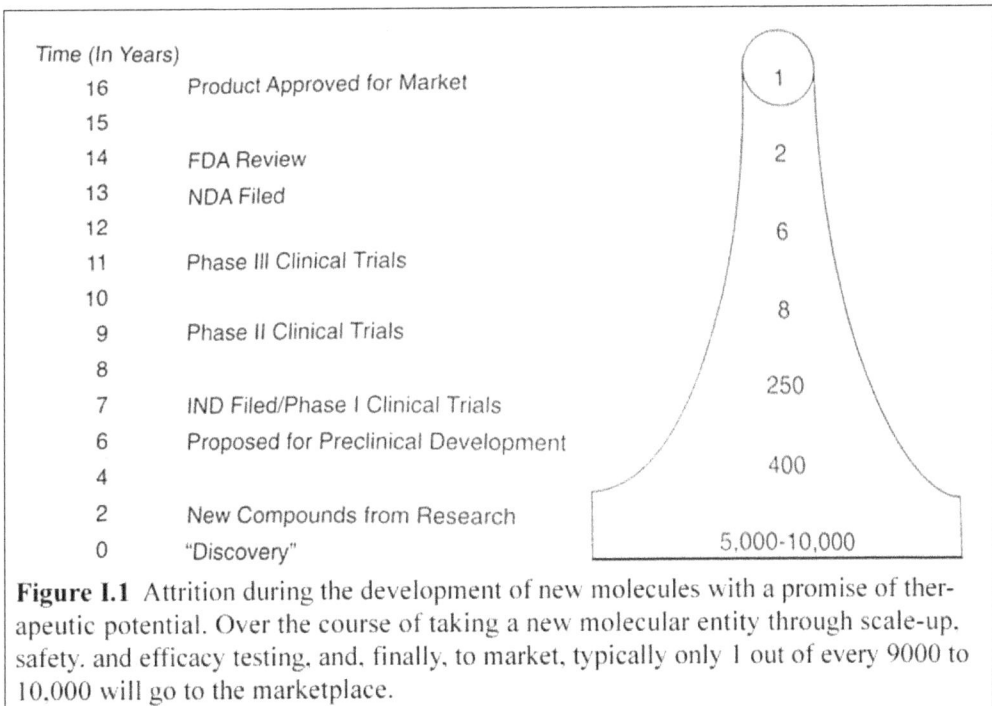

Figure 1 Pharmaceutical drug development pipeline (Gad, 2005, p. 2)

B. History of genomics in drug discovery

1. History of genomics

Genomics, the "systematic study of complete genomes," (Lander & Weinberg, 2000, p. 1780) first began to take shape in the 1980s as technical innovations began to allow for the identification of DNA polymorphisms. The field of genomics began to mature rapidly in the early 1990s in the footsteps of additional critical technical innovations in molecular biology and genomics: the first characterizations of human genes along with the perfection of the first PCR[1] machines. In conjunction with Oliver Smithies' invention of gel electrophoresis in 1955 (Smithies, 1955), such innovative techniques made feasible the rapid evaluation and determination of genetic variation between individuals. The Human Genome Project, completed in 2001, identified approximately 1.4 million single nucleotide polymorphisms (SNPs) in the 3 billion base pair-long human genome, with over 60,000 of them found in gene coding regions (Sachidanandam, Weissman, Schmidt, & et al, 2001).

2. Adoption of genomics in pharmaceutical industry

It is widely understood that different people respond differently to the same medication. In other words, variations within individuals give rise to variations in individual drug responses. Given that the study of the human genome at its root promised incredible detail of variation at the level of individual persons, it was perhaps inevitable that early interest in the adoption of genomics to pharmacological research would focus on identifying subsets of patients at special risk of adverse drug reactions (e.g., see (Breckenridge, 1996)). Just as the Human Genome Project helped build excitement over genomics, the development of genomics techniques led to the belief commonly held in the pharmaceutical industry that the power and promise of such techniques would materialize into a radical improvement of drug discovery. Genomic innovation and early adoption of genomic innovations seemed to promise a revolution in creating new drugs.

Pressures against the early adoption of genomics to pharmacology in large part appear to amount to two general problems: (a) the relatively high cost of computing resources, and (b) immaturity of genomic data and its uses.

In 1996, the cost of a single gigabyte (GB) of hard drive storage was approximately $250.00 (in 1996 dollars) (Smith, 2004) while in 2006 hard drive storage costs are as low as $.50 per GB. If we were to assume the entire human genome were sequenced and stored on a hard drive in 1996, a single instance of that 6GB of data would cost approximately $1500. While $1500 itself is not a prohibitive cost, the inclusion of each data set produced by analyses of the whole genome, typically within an order of magnitude of the object of study, would have cost hundreds of dollars to store. The costs of storing multiple data sets produced by analysis of the genome would quickly pile up. In short, it would have been possibly prohibitively expensive, just in hard drive storage alone, for a team of researchers within an organization to pursue information-intensive studies in 1996 working with a data object as large as the human genome. In addition to sheer data storage costs, the costs of computational processing power, computer memory, and expertise needed to manage such a data-intensive pursuit as genomics research would have been very high in 1996.

Such high computing and data storage costs would undoubtedly need to be justified by a high likelihood of significant returns. However, in 1996, sequencing of the human genome was not complete. Further, no one understood how to use the genomic data to identify targets or refine lead candidates. Completion of the sequencing of the human genome did not take place until 2001.

Sequencing the entire human genome, while a major accomplishment in its own right, did not directly provide information as to the functions of segments of that genome. We might best consider functional analysis as a process analogous to translating the Rosetta stone after the characters on the artifact were made visible. A sequence is, after all, not strictly speaking, a gene. The sequencing of the human genome in effect produced a sequence of base pairs, yet the full meaning of those sequences is not yet fully understood. Those meanings involve knowledge of a myriad of pieces: knowing which base pairs combine for RNA transcription, which base pairs merely hold physical positions that influence RNA transcription, which base pairs modify transcription, which base pairs are artifacts of evolution or lost transcription, and which base pairs do nothing at all. It is a story that tells no less than the *central dogma of molecular biology* itself: enzymatic proteins transcribe replicable DNA into RNA. Ribosomes translate RNA into proteins necessary for and present in nearly all biological function and variation including DNA replication (Crick, 1970).

The functional analysis of the entire human genome is only now underway, and the means for integrating the disparate parts of various functional analyses are only now being explored (Ekins, Bugrim, Nikolsky, & Nikolskaya, 2005). Functional analysis allows the researcher to bridge the gap between genotype and phenotype. One or more genes from an organism are specifically altered (typically removed or "knocked out") and the resulting phenotypic changes are observed. The observation leads to the knowledge of a gene's function in an organism.

C. Pharmacogenomics: a new model for discovery

1. Definition of pharmacogenomics

Since the early 1990s, drug companies have looked to genomics for innovations in drug discovery. The field known as pharmacogenomics was born out of the union of genomics and pharmacology in the wake of the completion of the Human Genome Project in 2001 (Altman & Klein, 2002). Pharmacogenomics (PGx) is the "study of the structure, content, and evolution" (Gibson & Muse, 2004, p. 1) of the human genome in order to help identify drug treatments for human disease. More precisely, PGx involves the "application of genomics information and technologies in drug discovery and development so as to identify, on the basis of genetic make-up, those individuals who will respond most favorably to a drug or those who are at risk of serious side-effects" (Oxagen Ltd., 2005). For example, such refinement helps narrow clinical trial populations, thereby reducing costs and increasing safety of clinical trials while increasing the chances that the NME will reach market (A. D. Roses, 2000). Pharmacogenomics not only helps researchers find better subgroup matches for specific drugs but also helps refine the ability to identify drug targets and genome-wide variation characteristics (A. D. Roses, 2000).

2. Pharmacogenomics versus "classical" drug discovery

> Drug discovery has become so complex that it cannot be contained within the
> confines of the pharmaceutical industry. Discovery and, for that matter, drug
> development need a diversified and flexible industrial base. (Drews, 2000, p.
> 1963)

Pharmacogenomics is fundamentally different from "classical" drug discovery, because

unlike classical practices, PGx depends on information-intensive practices (Ramanathan

& Davison, 2002). While the genetics-oriented technologies differ from older

technologies for drug development in many ways, the crucial difference is that PGx

techniques generate large pools of information requiring advanced analysis. For example,

a study showed that as of 1996 pharmaceutical therapies addressed approximately 500

molecular targets. Recent estimates place that number at upwards of 10,000 potential

targets (Drews, 2000). In particular, the advent and mass adoption of high throughput

screening (HTS) techniques (robotics that enable the automated simultaneous analysis of

large numbers of compounds) has made the production of large sets of data

commonplace. HTS tuned to the task of gene expression places data-intensive analysis

squarely in the middle of PGx efforts.

HTS represents only one of many ways in which informatics is intertwined with

pharmacogenomics. PGx, like most present biomedical research, is difficult to distinguish

from informatics. This crucial interrelation of informatics and biomedical research

differentiating PGx from classical drug discovery is commonly referred to as

bioinformatics, defined as

> a multifaceted discipline combining [...] computational biology, statistics, mathematics, molecular biology, and genetics [.... that] enables biomedical investigators to exploit existing and emerging computational technologies to seamlessly store, mine, retrieve, and analyze data from genomics and proteomics technologies [....] achieved by creating unified data models, standardizing data interfaces, developing structured vocabularies, generating new data visualization methods, and capturing detailed metadata that describes various aspects of the experimental design and analysis methods. (Fenstermacher, 2005, p. 440)

Just as information science might be seen as the interdisciplinary intersection of computer science, library science, mathematics, statistics, and cognitive science purposed for the human user, bioinformatics is likewise such an intersection but purposed towards a more specific type of user--the biomedical researcher.

On the relevance of bioinformatics Dr. Russ Altman, MD, PhD, of Stanford University writes,

> [b]iomedical informatics has gained prominence recently because biologists can now collect more data. The success of the genome sequencing projects has catalyzed a new way of thinking in biology, whereby data are collected on a large scale and without a particular hypothesis in mind. The data are then placed in a database, and scientists with hypotheses can extract information from the database in order to evaluate the merits of the hypotheses. This leads to a fundamental change in how some investigators do their work: Instead of first moving to the laboratory, they first move to the database, and only after assessment of the available data are experiments planned. (Altman & Klein, 2002, p. 114)

Given Altman's perspective on bioinformatics, it appears that PGx changes biomedical research from an effort characterized by the individual carefully building data points one by one at a lab bench to an effort where the researcher's most anecdotal and direct interaction with the subject is with the data itself. The database becomes the focal point of research. The impact of informatics is so fundamental that researchers design experiments based on large electronic data collections. PGx shifts the focus of drug research from data acquisition to data analysis.

3. Barriers to success of pharmacogenomics approaches

Pharmacogenomics appears to have yielded little fruit to date (Gad, 2005, p. 8). While the promise of increasing the throughput of the pharmaceutical pipeline has been high, it has led only to disappointment (A. D. Roses et al., 2005). Despite the appearance of genomics techniques over the last 20 years and the completion of the Human Genome Project in 2001, few if any drugs based on genomics research have reached the marketplace. While that may be partially explained by the length of time in which it takes a drug to travel the pipeline, it may also be understood by the lack of PGx-based drugs in the pipeline. For example, the world's second largest pharmaceutical company, GlaxoSmithKline, only started its first PGx-based drug target evaluation in March 2006. Drugs discovered using PGx have only now begun to enter the pipeline.

On the shortcomings of pharmacogenomics, Drews & Jurgen write:

> "It is difficult to judge the "success" of the new paradigm of drug discovery on the basis of published data. Some pharmaceutical companies have acknowledged that HTS has resulted in a large number of "hits"--an impression that is corroborated by a number of recent publications. However, some industry leaders have expressed disappointment that very few leads and development compounds, if any, can be credited to the new drug discovery paradigm. On the one hand, the meager results may be due to the relatively short period during which the new drug discovery paradigm has been seriously implemented. On the other hand, the lack of meaningful results may indicate that the system has not yet been optimized. What might have gone wrong during this initial phase?" (Drews, 2000, p. 1962)

Genomics research pioneer and Glaxo Smith Kline Senior VP for Genomics Research

Allen Roses has recently shed light on why pharmacogenomics-based approaches may

not be "optimal," as Drews & Jurgen put it. According to Roses, who arguably is in a

unique position to understand the problem, the central problem is one arising from

information struggles. Roses writes,

> What factors have limited target selection and drug discovery productivity?
> Although HTS technologies were successfully implemented and spectacular
> advances in mining chemical space have been made, the universe for selecting
> targets expanded, and in turn almost exploded with an inundation of information.
> Perhaps the best explanation for the initial modest success observed was the
> dramatic increase in the 'noise-to-signal' ratio, which led to a rise in the rate of
> attrition at considerable expense. The difficulty in making the translation from the
> identification of all genes to selecting specific disease-relevant targets for drug
> discovery was not realistically appreciated (A. D. Roses et al., 2005, p. 179).

What Roses calls the "noise-to-signal" ratio sounds like the problem of information

overload, yet it also sounds as if it borrows from the language of Information Theory as

put forth by Claude Shannon. Roses' insight seems to corroborate Ekins' observation that

already-extant data is not optimally utilized. Pharmacogenomics is failing to deliver

because PGx researchers and organizations utilizing PGx research have been unable to

meet the information challenges concomitant with the explosion of data.

Before I continue on to a discussion of information overload, let us first unpack the

features of barriers to pharmacogenomics success in light of concerns about low signal-

to-noise ratios. First, it must be noted that functional analysis is a new area. As reported

above, the functional analysis of the entire human genome has only just recently begun,

meaning that functional data is minimal in comparison to the totality of all genomic data.

Further, the means for integrating the disparate parts of various functional analyses are only now being explored (Ekins et al., 2005).

Another recognized barrier is the suboptimal use of already-accumulated preclinical and clinical experimental data (Ekins et al., 2005). The repurposing of old preclinical and clinical data for new PGx research is yet another under-explored area.

Another barrier to the success of pharmacogenomics relates to the concept of *genetic tractability*. While any of the 30,000 or so human genes can be considered a drug target, only a small percentage of them can realistically be considered as such. The difference is that not all 30,000 human genes can be acted upon or manipulated as of the time of writing this paper (November 2006). Only so many genes are presently manipulable. Only so many phenotypes can even in theory be specifically drugged to any effect. (To say that a gene is tractable is a little misleading; when we say a gene is tractable what we are usually saying is that we can control or manipulate the products for which it codes.) Further, while some genes may not be tractable at one specific moment in time, some genes have greater potentials for immanent tractability than others do. Finally, the issue of patent infringement remains. Even if a pharmaceutical company has identified a tractable target with characterized functionality, some aspect of either the target or the lead compound may be the property of another corporation or organization.

What exactly is the source of the lack of optimality that differentiates PGx from classical drug discovery? Pharmacogenomics is fundamentally different from classical drug discovery in that it is dependent on information-intensive practices (Ramanathan & Davison, 2002). How exactly does that dependence on information processing make pharmacogenomics suboptimal? In other words, which of the prominent barriers to PGx are information-based?

To sum, the list of prominent challenges to the success of pharmacogenomics are:

(a) Functional analysis is an underdeveloped study area,

(b) efficient integration of functional analysis data with genotypic data and clinical data is poorly understood,

(c) extant preclinical and clinical data are difficult to repurpose,

(d) identification of tractable targets is difficult, and,

(e) pursuit of drug research or development may be restricted by patents.

At least three of the five specific sources of pharmacogenomics' suboptimality listed above are primarily information management problems. The other two (nascence of functional analysis; identification of tractable targets) are at least in part problems of information management. Some type of information overload is, as GSK exec Allen Roses claimed, a central source of PGx's struggles to date.

IV. Pharmacogenomics-based Drug Discovery and Information Overload

A. Introduction

Pharmacogenomics experts have recognized that genomics-based approaches to drug discovery appear to suffer from some sort of information overload problem (A. D. Roses et al., 2005, p. 179). More specifically, the information explosion regarding the human genome may have been superseded by an explosion of noise leading to a significant attrition rate in the pharmaceutical pipeline (A. D. Roses et al., 2005, p. 179). However, it is not entirely clear how the concepts of information overload and signal-to-noise apply to information-based struggles in pharmacogenomics. In order to improve our understanding of the barriers to optimal use of pharmacogenomics information for drug discovery purposes we must first briefly unpack competing ideas about information overload and signal-to-noise and then contextualize the appropriate ideas within PGx-based drug discovery (PGx-DD).

B. How do we explain "too much information" in PGx-based drug discovery: Information Theory or information overload?

As shown in the previous chapter, the language Allen Roses uses to describe struggles with information in the field of PGx-based drug discovery refers both to a *signal-to-noise* ratio and to information overload. The terminology appears, however, to be rather ambiguously utilized in the context of PGx-DD. "Noise-to-signal" seems to refer to

Claude Shannon's mathematical theory of communication (Shannon & Weaver, 1949)

while the problems described by PGx professionals sound more like cognitive issues

related to more formal notions of information overload.

1. Shannon's Mathematical Theory of Communication

In 1948, Claude Shannon of Bell Labs completed work on his mathematical theory of

communication. In so doing, Shannon is credited as fathering the field of Information

Theory. It is from Shannon's theory that the notion of *signal-to-noise* arises, among many

other concepts crucial to any understanding of information. In his introduction to the

ensuing book publication comprising Shannon's work on the theory, Warren Weaver

explains that the theory was purposed to deal with three distinct levels of

communications problems, as follows:

> Level A. How accurately can the symbols of communication be transmitted? (The
> technical problem.)
> Level B. How precisely do the transmitted symbols convey the desired meaning?
> (The semantic problem.)
> Level C. How effectively does the received meaning affect conduct in the desired
> way? (The effectiveness problem.) (Shannon & Weaver, 1949, p. 4)

Information in Shannon's sense is not used in the ordinary sense of *information*. While

by *information* we ordinarily mean something akin to what has been said, Shannon means

it in the sense of what may possibly be said (Shannon & Weaver, 1949, p. 8). For

Shannon, information is a probable message sent over a channel (e.g., a telephone wire)

and his concern is with describing general properties of the transmission and

interpretation of such electronic signals.

Concerns about the ratio of signal-to-noise with respect to information transmission do originate from Shannon's own communication theory work. The very ratio of signal-to-noise appears in Shannon's own theoretical examination of channel capacity with power limitation (Shannon & Weaver, 1949, p. 100). Shannon uses the ratio of the power source of the signal (denoted as P) to the power of the noise (denoted as N) in order to provide a general way of calculating how many bits per second any communication pathway can actually transmit. Shannon replaces P with S, the peak allowed transmitter power, in order to adjust channel capacity where peak power limits the rate of the channel to transmit bits. According to Shannon the upper bound rate of a channel is the channel band times the log of the ratio of signal plus noise to noise where the signal-to-noise ratio is low (Shannon & Weaver, 1949, p. 107). Loosely speaking, the rate at which telephone wires, coaxial cables, wireless networks, and the like can transmit messages varies logarithmically with the ratio of peak power (signal) to background noise on the channel (noise).

Shannon's specified problem set does not accurately match the sort of problem a drug discovery researcher is facing, not at least without a considerable stretch. Shannon's sense of *information* in his definitive work on communication theory does not seem quite the same as the sort of information we are dealing with when we speak of genomics research data. Finally, Shannon's notion of signal-to-noise can at best only loosely apply to notions of researchers struggling with too much information in their hands. Shannon is writing about communication channels, not people.

Any effort Shannon may have made to model human communication in his theoretical work was at best tertiary to the central thrust of his work, which was to generalize the properties of electronic communications systems. In short, Information Theory as proffered by Shannon does not appear to apply in a straightforward way to the sort of "noise-to-signal" problem Allen Roses describes. The signal-to-noise problem Roses reports is an information problem but it appears to be an information problem unlikely to be either explained or resolved through the lens of Shannon's communication theory.

2. Information overload

The concept of the possibility of too much information dates back to ancient times (Bawden, Holtham, & Courtney, 1999, p. 249). The recurring concern of *information overload* stems from the general notion that a person's work becomes inefficient from increasing difficulty experienced in locating the best pieces of information. With the advent of computer-based information retrieval systems in the 1950s (Bawden et al., 1999, p. 249) as well as the beginnings of the mass proliferation of scientific research literature (Ziman, 1980), the concern became more frequently and more directly articulated and investigated. While any exact definition of information overload is elusive, issues of relevance and efficiency are commonly noted, as are issues of both data management and psychic strain (Bawden et al., 1999, p. 250). The constant problem however is that *information overload* stands for a struggle—a struggle that increases as a collection of information grows beyond human tractability. The recurring solution inevitably takes the form of methods or techniques that allow a person to locate some tractable set of pieces of information of sufficient quality in a reasonable amount of time in order to aid the person in completing the task.

C. Impact of information overload on PGx-based drug discovery

Information overload describes the general problem of "noise-to-signal" referred to by Allen Roses. Roses characterizes the information problem facing PGx-DD as having increased the rate of attrition of drug candidates in the pharmaceutical pipeline. Further, he states that the solution to the problem is an increase in "specific, disease-relevant targets" relative to all genomic data (A. D. Roses et al., 2005, p. 179). In other words, the proliferation of genomic data has drowned out this highly specific disease-relevant genomic information to the point that it increases drug discovery failure. The way to resolve the issue is to reduce information overload in PGx-DD by restricting the flow of information to PGx researchers to highly specific disease-relevant genomic information. As Roses says, providing researchers with validating evidence is crucial.

What, however, frames, delimits, or describes *validating evidence* for candidate targets? Roses states that disease-specific targets chosen based on well-trod beliefs "have a significant probability of being the totally wrong target" (A. D. Roses et al., 2005, p. 180). It is therefore not enough to identify highly specific disease relevant data efficiently; the data must support infrequent or entirely novel theories. The data must in essence have the characteristic of supporting *novelty*, of supporting ideas not commonly held, of bolstering theories that appear to be unreasonable.

The quality of data for PGx information should be evaluated on the following three criteria:

(a) the disease-relevance of the information,

(b) the specificity of the information, and

(c) the novelty of the information or the novelty of the theory supported by the information.

V. Text Mining: An Optimal Information Overload-reducing Technology For PGx-
   based Drug Discovery

   A. Chapter overview

The present chapter aims to illustrate text mining and its uses. The two most prominent

goals of this chapter are the establishment of a novel explanatory framework for text

mining and a cogent assessment of text mining's value to PGx-based drug discovery

using the framework. Each requires a number of sequential steps to be taken before each

is established.

Developing a new text mining framework will require several crucial elements to be

postulated and supported. First, a brief definition of text mining will be offered. The

novel concept of *information overload-reducing technologies* will be defined in brief.

Text mining will be compared with similar technology types such as information

extraction, data mining, and information retrieval. With the brief definitions in hand, a set

of categorical attributes and values will be defined and employed to describe text mining,

information extraction, data mining, and information retrieval, resulting in a convenient

comparison grid.

Having an in-depth picture of text mining and information overload reduction will help in describing successful applications of text mining with an eye towards PGx. Success stories of text mining applications in the business world will be reviewed, followed by a review of successful text mining applications in the biomedical domain.

The discussion of text mining will culminate in an examination of how text mining may reduce information overload for PGx-DD. The present chapter aims to establish text mining's potential and actual value relative to the overlapping tasks of reducing information overload and improving PGx-DD.

B. Definition of text mining

Text mining, according to leading researcher Marti Hearst, is "the discovery by computer of new, previously unknown information, by automatically extracting information from different written resources" (Hearst, 2003, ¶ 1). Text mining is frequently first defined by differentiating it from data mining and then is further defined by how it differs from other information processing techniques (Hearst, 2004; Weiss, Indurkhya, Zhang, & Damerau, 2004).

The following sections will take an approach to defining text mining similar to Hearst's. However, before a more exhaustive approach is completed a succinct definition may suffice. *Text mining* is the automated derivation of novel information from extant texts.

C. Information overload-reducing technologies

A number of types of information management systems help their users handle large quantities of information otherwise too overwhelming to use. Information retrieval (IR), information extraction (IE), data mining (DM), and text mining (TM) tools all help users reduce the strain created when confronted with too much information. For the purpose of the present study, the set of tools falling under the four aforementioned general categories will be collectively referred to as *information overload-reducing* technologies. IR, IE, DM, and TM do not exhaust the possibilities for information overload-reducing system types.

The four aforementioned types have been chosen because they are types often referred to as similar or related technologies. Text mining is sometimes referred to as a type of information retrieval; information extraction is sometimes referred to as text mining, and text mining is frequently referred to as a type of data mining. Referring to all of them as *information overload-reducing technologies* and subsequently enumerating crucial differences between them may seem tedious at first. I believe that such distinctions will serve to highlight the particular strengths of text mining for pharmacogenomics-based drug discovery.

D. A comparison of text mining to information retrieval, information extraction, and data mining

1. General commonalities

Text mining is a hybrid form of the more mature fields of *information retrieval, information extraction,* and *data mining* (often referred to as *machine learning*). Like information retrieval, text mining is primarily text-centric and aims to reduce information overload pressures by finding highly relevant texts from large text collections. Like information extraction, text mining uses pieces of text excised from files and other larger data structures. Text mining, like both information retrieval and information extraction, often makes heavy use of computational linguistics, particularly in the process of structuring the text data. Like data mining, text mining utilizes statistical learning to identify or generate useful patterns from the input data.

2. A detailed comparison of TM, IR, IE, and DM

a. Purpose of the comparison

The purpose of the following comparison of information overload-reducing technologies is to help distinguish text mining from other information overload-reducing applications. While many different definitions of text mining abound most are defined in an informal way. Such informal definitions make it difficult to distinguish true text mining applications from applications mislabeled as text mining. The descriptive and slightly more formal framework will help us not only to characterize extant applications but also to maintain a finer granularity for examining similarities and differences. Finally, such a descriptive model affords us the ability to make a better assessment of the future role of text mining applications and to relate those projections to known problems in PGx-DD.

Typically we might see these four types of technologies all represented as information retrieval tools. We might see information extraction reported as a type of text mining, and text mining as a type of data mining. Moreover, we might see that grouping considered as being part of a greater group of information retrieval tools. However, information retrieval in its most basic sense is literally one of retrieving, of finding. Text mining and data mining systems do not perform retrieval for the end user. What they are doing is creating new material rather that recalling existing material. Text mining and data mining alike learn from the existing material and present new information synthesized using the existing material as input. Further, while information extraction is commonly seen as a particular type of text mining (see Hearst, 2004, or Weiss et al., 2004, for examples), it essentially employs advanced finding techniques just as information retrieval systems do. The present characterization of the four system types as separate subtypes of information overload-reducing systems is a novel characterization unique to the present study, and so too is the distinction between information extraction and text mining.

The general spirit of the description is to identify features of the four differing types of systems along axes of general information system characteristics: *input, output, processing*, and *use*. A secondary motivation is to elucidate the notion of hypothesis formation, namely, how it is an essential quality of all four technologies rather than merely a special case of text mining.

      b. Comparison features

        i. Use

          1. General problem-solving task

The problem-solving attribute answers the general question, "what is the aim of using the application?" With the typical case of information retrieval, such as a search engine, generally the user aims to find some document relative to their information-seeking needs. With information extraction, the search behavior is finer-grained, as the hope is to extract relevant words, phrases sentences, and/or paragraphs from the documents. As for data mining, the general goal is to derive previously unknown and undocumented relationships such as correlations from a structured data set, usually stored in a database. Like data mining, text mining also aims to learn something previously unstated, unlike either information extraction or information retrieval. However, unlike data mining, text mining starts with unstructured text as input, just as IR and IE do.

ii. Input

The general paradigm of computational systems must always be described in terms of input, processing, and output. Solely relying on user intent (*i.e.*, problem-solving task) leaves our categorization incomplete. It should not therefore be surprising that if I can identify groups of applications by distinctions about input, processing, and output then I can indeed rely on these distinctions to classify them. Input can be differentiated and categorized along three subdimensions: *input type*, *input content*, and *input structure.*

1. Type

Input type describes what sort of actual item is being put into the computational system. Are we putting files, strings, database records, references, etc. into the system for processing? Another way to ask the question is to ask what sorts of things are in the pile of things that are making us suffer from information overload. Unsurprisingly, text mining shares characteristics of DM, IR, and IE in this respect. Like DM, TM often takes db records as input, though most of the time TM, like IR and IE, takes files as input.

2. Content

When we say *content*, we are referring to what is contained in the input (the input type) that is being analyzed. The content put into TM systems, like IE and IR, is primarily text. Database records processed by DM systems typically contain numeric, cardinal, or ordinal fields. It should be noted that data mining often has text input, but the crucial difference is that the input for data mining usually is limited to single words, phrases, and names rather than sentences and paragraphs.

3. Structure

Text is usually described by computer and information scientists as *unstructured*. To a poet or English teacher text may seem highly organized, having distinct formal features. However, what we mean by *structured* is that the data is contained in a *data structure* in the computer science-specific sense of the term. Text must be arranged in logical relationships suited for efficient computational processing before it can be utilized by an IR, IE, or TM system. However, the database records used as input for a DM system are already organized in such a fashion.

iii. Processing

1. Role of database

In information overload-reducing systems, the database frequently has an important if not central role. While the role of a database to a DM system is primary—as input—the role of a db to the other three types of systems is intermediary. Often the input for IE, IR, and TM systems is given structure and stored in a database. Input can be processed within a database. Alternately, it can be retrieved from a database, processed outside of that database, and subsequently returned to the database.

2. Roles of pattern learning and pattern extraction

Data mining always involves statistical learning while information retrieval and information extraction usually do not. It is a conceit of the present study that text mining involves statistical learning. By making this assumption, I am able to categorize these systems in such a way as to give the notion of usefulness primacy. In particular the crucial difference to be made here is whether the role of the system is to find something already known, stated, stored (as with IR and IE) or to come up with something novel. Statistical learning is the primary means by which text mining systems produce novel results. As the present study will show, novelty is a central concern of PGx researchers that is under-realized in part by a general failure among PGx researchers to differentiate finding/matching/extracting tasks from text pattern learning/derivation/discovery tasks. Pattern learning is a means to novelty.

iv. Output

1. Type

As with input, *output type* describes the things that the system produces. IR returns files usually with some sort of summary page tying the files together. Like IE, TM systems often return pieces of text usually at the word or phrasal level. However, like DM, TM systems can also return rules that describe patterns or relations across different attributes.

2. Novelty

Simply put, the task of a mining system, whether structured data or unstructured text, is to arrive at a novel pattern. IE and IR systems do the work of finding as opposed to discovering or deriving novel information. The notion of *discovery* can be misleading, as, for example, some would argue that Christopher Columbus discovered America when in fact he merely had found it where it had always been (and already inhabited by other people). For present purposes, I will avoid the word *discovery* as best as possible; the concept of *deriving novel information* will take precedence. Further, a strict understanding of *novelty* and *discovery* will help us assess how to use text mining to assist PGx professionals working towards *drug discovery* and *pharmaceutical innovation*.

For the purposes of the present study, *novel information* is operationally defined as information that is new by virtue of the system itself and cannot be found anywhere within the input of the system. Novel information is information generated by the system that did not exist before the system processed the input. For the present purposes, I will not attempt to include such subjective notions as *interestingness* or *surprisingness* into the definition of *novelty*. (Notions of *quality* tuned to the particular task of PGx-DD, however, will be discussed in the final section of the present chapter).

3. Found vs. derived

It is important to make a crucial distinction between *finding* and *deriving* when discussing information overload-reducing application classes. The object of IR and IE is to find things amidst a large collection of documents; the object of data mining is to derive via statistical learning a pattern given the input. Text mining likewise derives patterns but does so for IR- and IE-like inputs.

4. Factual or hypothetical

It is crucial to first stress that the *factual or hypothetical* attribute does not allow me to distinguish between the application categories described herein. Usually hypothesis generation is associated with a particular class of text mining applications. Upon reflection, however, the primary function of all four classes is to produce not a fact but rather a hypothesis. That is to say, the factual basis for any information retrieved, extracted, or derived is always external to the system. Everything produced by any overload-reducing application is necessarily subject to further review and verification by a user. Even in the simplest of IR tasks, such as an internet search, the relevance of the results given the user intent must be measured by the assent or dissent of the user. In other words, the set of search results as a whole supports the hypothesis that at least one search result is relevant to a user given the user-provided search terms.

5. Verification

Given that all results produced by all four of the system classes described herein are hypothetical, any valuable or constructive use of the results require some form of *verification*, particularly for the purposes of research. In IR, the user must at least check whether the documents deemed highly relevant by the system are indeed relevant to his/her intentions. The ultimate relevance judgment rests with the user. Likewise, extracted statements from IE systems and patterns derived by TM and DM systems also beg for external verification. The results of IE, IR, and TM systems cannot be judged true solely based on the input alone. However, with DM systems to some extent, given that the input is already structured, the patterns can be at least internally validated with statistical techniques such as cross-validation. While cross-validation can be employed with TM-based statistical learning as well, the cross-validation provides no information about the validity of the results given the user's intent. Perhaps the difference can be explained by the notion that the user's intent is *prima facie* fully described by the input. Data mining input usually is not considered a structured representation of unstructured phenomena but rather structured information.

c. Information overload-reducing systems comparison grid

Attribute type	Attribute	IR	IE	TM	DM
Use	**Problem-solving task**	Find document about X	Find text in document about X	Learn something new from text	Learn something new from db
Input	**Input type**	Files	Files	Files and/or db records	Db records
Input	**Input content**	Text, multimedia	Text (may contain table elements or images)	Text	Numeric, cardinal, or ordinal data fields
Input	**Input structured**	No	No	No	Yes
Input or Processing	**Typical role of db**	Intermediary	Intermediary	Input or intermediary	Input
Processing	**Involves pattern learning**	No	No	Yes	Yes
Processing	**Involves pattern extraction**	Yes	Yes	Sometimes	No
Output	**Output type**	File	Text	Text and/or rule	Rule
Output	**Output novel**	No	No	Yes	Yes
Output	**Output found or derived**	Found	Found	Derived	Derived
Output / Use	**Output factual or hypothetical**	Hypothetical	Hypothetical	Hypothetical	Hypothetical
Output / Use	**Verification locus of hypothetical output**	External only (user-based)	External only (user-based)	External only (user-based)	Internal (e.g., cross-validation) and external

Table 1 Categorization of Information overload-reducing system classes

3. Types of text mining

Three general types of methodologies and practices dominate text mining: *automatic text categorization (text classification)*, *clustering* and *relationship derivation*. Each type utilizes similar techniques and technologies, yet the merits and purposes of each differ significantly. Customarily, text categorization, clustering, summarization, and information extraction are considered species of text mining (*e.g.*, see Weiss et al., 2004). However, summarization, referred to by its creator as "automatic abstracting" (Luhn, 1958), is a species of information extraction (Swanson, 1988a) rather than an application designed for derivation and synthesis. While summarization systems may incorporate methods for generating new information, in general it is more extractive than generative.

a. Automatic text categorization

Automatic text categorization, often referred to as *text classification*, is typically used to address one of two deeply interrelated problems. Either it is used to (a) identify the category to which new documents should be assigned based on previous manual categorization of other documents in the collection, or (b) forecast events based on a previously established but hidden associative trends between extant documents and earlier states in those associative trends. Automatic text classification is heavily dependent on the data mining paradigm: some machine learning algorithm evaluates a feature representation of a training set of previously categorized documents and tries to determine to what category new documents belong. The particular type of pattern learning utilized in automatic text categorization is referred to as *supervised learning*. The most common use of automatic categorization is for binary category assignments— whether a given item belongs or does not belong to a predefined category. Categorization and classification models of greater complexity can be constructed using combinations of binary category assignments. However, the process of automated category assignment can handle larger number of classes.

For example, ibiblio.org (http://ibiblio.org) is host to approximately two thousand independent web collections. Many of the collections are catalogued on the ibiblio.org site according to the Universal Decimal Classification. A large number of the sites in the collection are not classified, however, and a growing number of new collections are added every day. The manual task of categorizing each site would be too time-consuming given the organization's small staff and the rate of new collection additions. Yet using machine learning to devise a model for determining the category of new sites would epitomize a good use of automated text classification techniques.

The automatic categorization of previously unclassified documents based on the previous manual classification of other documents leads us naturally to the ability of text classification (text categorization) to aid in forecasting. Instead of associating some characteristic of a document with the document as with the previous example, forecasting through text classification aims to associate an external event with the existence of a set of documents. Imagine for a moment a set of documents from old news reports and press releases on the subject of sugar farming that are assumed associated with either increases or decreases (our two classes) in the price of sugar as a commodity. It is then easy to imagine that using the old documents to devise a model to make forecasts about sugar prices as new reports materialize is not only possible but quite useful and, fortunately, largely unexplored. The foundation for such a use of text classification is quite reasonable, as data mining for predicting stocks and commodities pricing has remained in vogue for years; the structured data used in data mining practices is frequently derived from unstructured text (Mittermayer, 2004, p. 1).

While the task of automatic document categorization might appear on the surface to be merely some sort of simple information overload-reducing task, it is important to note that the task of using documents to forecast externalities is strongly deductive and inferential. Such a task is different from merely reducing the number of items in one's pile; the task is equivalent to making that pile tell you the answer to the very question you are asking even if that answer is not literally contained within that pile. Further, automatic categorization strategies that look instead at passages such as paragraphs and sections within those documents (*e.g.*, see Theeramunkong, 2004) promise to push us further from the realm of simple information overload reduction into more advanced modes of deduction, inference, and synthesis.

b. Clustering

Text categorization makes use of supervised learning. The type of machine learning used in text categorization is referred to as *supervised* because the learning algorithm is given examples of class inclusion and exclusion that serve as the basis for learning a categorization model and making subsequent category assignments based on that derived model. Clustering, however, embodies an *unsupervised learning* approach. It addresses the question, "what do we do when we have no previous evidence of any classes or groupings whatsoever?"

Clustering algorithms learn by measuring highly complex distances between individual documents in a highly dimensional mathematical space for the purposes of finding feature clusters. The items clustered may be documents, words, topics, or more complex features. For example, I might take a number of stories from the news wire on a single day, cluster them, and see that the documents clustered together might have quite a bit in common. One grouping may share many terms in common referring to "heat", "weather", "record", and "temperature," thus indicating a number of documents focusing on the record temperatures set in some city. Another cluster may contain many terms such as *Hurricanes, NHL, contract,* and *salary cap,* thus indicating a set of document containing news about the Carolina Hurricanes professional hockey team.

While clustering has many useful applications, the efficacy of clustering suffers from the fact that there is no such thing as a *naturalistic* document cluster—the distances between documents depend entirely upon subjective decisions regarding notions of closeness as well as choices about what features to use. No one grouping scheme of a set of documents is more inherently true of that set of documents than another scheme. Feature representations for clustering, such as individual words, are often rudimentary; what makes two documents similar or different is a highly variable decision subject to opinion; and most importantly, the number of groupings desired can highly influence the groupings (Herron, 2005). Further, the evaluative measures commonly used in clustering research have been taken directly from data mining and from information retrieval without being appropriately rewritten for cluster analysis (Herron, 2005). Specifying in advance subjective preferences for evaluating the results is essential for the results to have meaning (Herron, 2005). With the subjective evaluative preferences articulated in advance, clustering results can be just as meaningful as results from classification methods.

Classification and clustering provide complimentary means for performing content analysis of large document collections. Clustering can help someone choose basic classes, and classification can then build upon the groups suggested by clustering.

c. Relationship derivation

Relationship derivation applications generate pieces of information derived from multiple sources. No one source is sufficient for such relationships; they must be derived from a minimum of two sources via the application of some transformation rule or evaluative procedure. The goal of relationship derivation is to discover a previously unknown relationship between two things or phenomena given a corpus. Relationship derivation may partake of many different technologies including machine learning, information extraction, and *inductive logic programming (ILP)*.

Oren Etzioni, Director of the Turing Center at the University of Washington, leads a comprehensive research project known as KnowItAll. The KnowItAll project focuses on the use of what the project team calls *unsupervised information extraction* for constructing automated machine reading tools. *Machine Reading* (MR) is different from IE and question answering (QA) in that while "IE and QA focus on isolated 'nuggets' obtained from text [...] MR is about forging and updating connections between beliefs."

47

(Etzioni, Banko, & Cafarella, 2006, p. 1) The operating principle of MR is basic yet ambitious; Etzioni cites Roger Schank's example that if a text says a person has just left a restaurant after a satisfying meal, "it is reasonable to infer that he is likely to have paid the bill and left a tip." (Etzioni et al., 2006, p. 1) The KnowItAll project has led to the construction of OPINE, a tool for mining consumer product reviews. While OPINE is still in development, a demo version is available online (http://knowitall-1.cs.washington.edu/Opine/Search.aspx). The MR project promises numerous relationship derivation applications.

A similar approach to MR, albeit a supervised one, has been taken by Stephen Muggleton of the University of London, Imperial College. Muggleton has helped pioneer the use of machine learning for inductive logic programming. Muggleton describes ILP as a process of explicit hypothesis generation and "knowledge discovery" (Muggleton, 1999). By definition, ILP is used to generate statements from a structured pool of data containing past examples, knowledge sources. The programmatic approach is based on first order predicate logic, particularly inductive hypothesis formation techniques (Muggleton, 2003). Unsurprisingly Muggleton has turned his attention to the domain of pharmaceutical drug discovery, citing previous success with ILP for highway traffic data analysis (Muggleton, 1999).

The present description of relationship derivation may seem somewhat vague. The vagueness is present namely for three reasons: relationship derivation is based on concepts considerably more complex than those guiding text categorization and clustering; relationship derivation is a nascent, largely undeveloped area difficult to name and describe; and finally, in a later section science-specific applications will be discussed in which relationship derivation will be the focus.

4. Examples of successful text mining applications

a. Business intelligence applications of text mining

While developing even the most general understanding of text mining seems an abstract affair, text mining is rooted in and regularly used for real-world tasks in the business community for various management efforts. The best uses for text mining in managing business affairs currently involve engineering enterprise content management for improved forecasting and customer relationship management (CRM).

Organizations are awash in texts, and the information contained therein is under- or un-utilized. From spam, to meeting notes, to a competitor's white papers, to industry news stories, employee reviews, and even to free-text customer complaints, organizations possess text that, if properly managed and mined, can lead to improved trend detection, spam filtering, operations decision-making, and responsiveness to both employee and customer needs.

i. Forecasting and enterprise content management

As discussed in a previous section, text classification can reasonably be used to make forecasts about many different things. Developers at Fireman's Fund Insurance Company were able to make use of adjuster notes and other free-form text entry fields in order to enhance detection of fraudulent auto insurance claims as well as prediction of deleterious homeowner claim trends (Ellingsworth & Sullivan, 2003).

In the Fireman's Fund Insurance Company example, a large number of fraudulent auto claims contained comments from insurance adjusters containing claimant references to Loss Adjustment Expenses (LAE). Fraudulent claimants seem to have a rather unusual familiarity with insurance industry terminology, particularly in conjunction with other telltale signs, such as frequent use of the word, "anxious."

Fireman's Fund Insurance Company noticed a massive growth of homeowner claims in one state in a brief period. Upon further inspection, it appeared that mold was almost entirely responsible for the increase. Fireman's Fund used past records of mold claims to head off advancing future mold claims, notably by detecting mold claims that presaged more serious and costly mold claims.

While the use of text mining for making market trend predictions has not reached its fullest maturity, it is evident from the Fireman's Fund example that text mining for forecasting is already a commercially viable and advantageous methodology for augmenting the management of core business practices.

ii. Customer relationship management

Customer Relationship Management (CRM) is a term that encompasses the total efforts embodied in the applications and methods an organization utilizes in order to manage relationships with its clients. Just as we saw with the forecasting example, organizations can and do use text mining in order to improve core business practices, particularly with the way they respond to customer needs. Randall Collica, a senior business analyst at HP, recently made use of text mining techniques and tools in order to improve customer relations in two crucial areas (Collica, 2003).

HP made use of SAS's Text Miner software package in order to improve the organization's responsiveness to telesales and overall product line purchasing. Representatives in HP's inside sales & call centers regularly make use of free-form text fields when conversing with customers. The SAS Text Miner tool enabled Collica to be able to tap into the customer text data by applying various categorization schemes for further evaluation and enhancement of customer service. Further, HP had continually struggled with products it acquired through corporate mergers and acquisitions (e.g., HP acquired Compaq in 2002) due to the fact that data about acquired products were either in structures not in line with HP products or simply not structured at all. Collica was able to utilize past invoices to identify the product line (e.g., desktop, printer, etc.) to which each individual legacy product belonged. Collica and his team were able to classify with 90% accuracy over 1 million previously unclassified parts.

   b. Biomedical discovery applications
       i. Introduction

The purpose of text mining systems is to discover novel information using large text collections as input. Many of the early and best examples of text mining systems in the true sense come in the form of biomedical-specific applications. Before I begin a discussion of PGx-DD-specific applications, I would like to touch on some successful text mining applications specifically designed to reduce information overload in the biomedical domain.

ii. Arrowsmith

In the 1950s, University of Chicago library science researcher Don Swanson began a career focusing on the development of computationally viable solutions to the problem of fragmented knowledge in the sciences (Swanson, 1960). After literally being struck by lighting in 1985, Swanson made a realization while doing some medical research. He noticed that two separate medical research papers that did not refer to or cite each other provided an answer to a question neither paper could help answer independently. After the 1986 publication of the seminal article, "Undisclosed Public Knowledge" (Swanson, 1986b) Swanson went on to develop a methodology for discovering what he termed as "complementary but disjoint structures in the literature of science" (Swanson, 2001).

In short, Swanson showed that multiple papers could be put together in a logical and procedural way to help the user to formulate novel scientific hypotheses. His method identified what I term *topical transitivity*: establishing a linkage using topics (terms, concepts, entities, subjects, phenomenae, etc.) across otherwise-unrelated documents helps conjoin literatures to aid in the formulation of novel hypotheses. In particular the literature can be utilized to associate two ideas that have never been directly associated.

The literature-based topical transitivity relation Swanson utilizes operates in the following way. If we are looking for an association along with supporting literature of the association between aardvarks and coffee, we could establish it by first exploring the topic set implied by all articles relevant to aardvarks as well as the topics relevant to coffee, then searching for documents containing topics associated with both *aardvark* and *coffee*. If we find a document substantiating the claim that aardvarks like to eat beans and another document stating coffee is a type of bean, then we have a literature that supports some tentative connection between aardvarks and coffee.

After describing and developing the method of conjoining disjoint literatures for hypothesis formation, Swanson went on to illustrate the efficacy of his idea with several solid relations in the medical literature. Among those relations Swanson discovered: (a) overtraining & resulting inflammation factor in atrial fibrillation (Swanson, 2006); (b) causal relationship between magnesium levels and migraines (Swanson, 1988b); and therapeutic relationships between (c) Fish Oil and Raynaud's Syndrome (Swanson, 1986a), and between (d) arginine and degenerative diseases marked by low levels of somatomedin C (Swanson, 1990).

Swanson's earliest stated goals for resolving the problem of information overload have included making such solutions programmatic and computationally viable (Swanson, 1960). After developing his method and finding supporting examples (heavily supported by his intensive use of databases), he began to work with University of Illinois psychiatry researcher Neil Smalheiser to develop a computer application to help a user discover such relationships in the medical literature. Named *Arrowsmith* (http://arrowsmith.psych.uic.edu/cgi-bin/arrowsmith_uic/start.cgi), the now-online tool allows the user to find complementary but disjoint literatures in PubMed and hence formulate novel hypotheses[2].

The power of Arrowsmith is that it can help a researcher easily and efficiently leverage the massive body of literature contained in PubMed in order to generate a novel scientific hypotheses supported by research literature. While Arrowsmith does not generate any hypotheses itself, it is powerfully augmentative in that it points a clear path for a researcher to explore, analyze, and articulate novel and highly specific relations between otherwise distinct scientific findings.

iii. PubMiner/BioPubMiner

PubMiner is a machine-learning based text mining system designed to mine MEDLINE (Eom & Zhang, 2004a, Eom & Zhang, 2004b). PubMiner discovers relations across the literature between genes and gene products and provides the user with a detailed analysis of the specific relations. The analysis includes the relevant literature that empowers the user to investigate the system's recommendation further. A key feature of the system is that it integrates public genomic databases such as the *Saccharomyces* Genome Database (SGD) and the Munich Information Center for Protein Sequences database (MIPS). The system makes use of two learning techniques. The Apriori association rule finder (Agrawal, Imielinski, & Swami, 1993) discovers interactions among feature sets, and a clustering algorithm evaluates the results of the rule finder to uncover feature distributions and provide information needed for hypothesis formation. The hypothesis is shown to the user as a complex network of relations between biological entities drawn as a graph along with the relevant extracted text for the nodes and relations displayed in the graph.

While PubMiner depends on supervised learning its precision scores for thousands of candidate interactions was nearly 95%. The authors cite the 5% gap may due to the relative high frequency of false positive data generated by high throughput data informing the public databases used. Mitigating the value of the tool is the amount of supervision needed to produce its relational networks though to what degree is unclear.

iv. Robot scientist: soup-to-nuts discovery

ILP creator Stephen Muggleton (see section C.2.c.3 of chapter V) collaborated on a team project to build a robot scientist. Ambitious in scope, the collaborative undertaking of the British scientists produced a closed-loop robotic system. The robotic system mines research data, formulates hypotheses, and subsequently designs and executes experiments to test those hypotheses. The system then uses the new findings as further input for future hypothesis construction and experimentation (King et al., 2004). The robot was assigned the task of determining gene function in yeast via high throughput methods typical of core PGx research practices. What was remarkable was that the system, rather than a scientist, successfully devised hypotheses by reading the research data and literature, designing and executing experiments, and integrating the large amount of data generated from each high throughput experiment into a periodic evolving cycle of hypothesis-experiment-analysis.

The robot has been nothing less than a remarkable success. The heart of the robot scientist's success is its hypothesis generation engine. One of Muggleton's ILP systems, named Progol, acts as the robot's brain for the task of formulating hypotheses. Progol uses background knowledge as input for a set of well-defined inductive logic procedures. (An example of inductive logic, in particular inverse deduction, might be deducing that all swans are white from a data set consisting of objects such that all the swan members are white.) The product of such procedures is at worst a good approximation of scientific reasoning and its efficacy is demonstrated in part by the robot's performance. The robot has been demonstrated to perform at least as well as the graduate students who would typically be responsible for the tedious yet important task of characterizing gene function for thousands of yeast genes. While the rules of deductive inference are thousands of years old, they were only made algebraic in the 19th century, and inversion of deduction was devised as a means for inductive inference for the first time in the late 1980s (Muggleton, 1990).

E. How text mining can reduce information overload in PGx-based drug discovery

As evidenced by the robot scientist, adaptive learning technologies can be utilized to overcome PGx-specific overload problems in a comprehensive and fully automated fashion. Muggleton's research has already extended the potential for the robot scientist to go beyond the limited knowledge database into supervised learning of literature databases. Far less manually curated approaches to devising new hypotheses from the biomedical research literature such as Arrowsmith and PubMiner have shown to be successful for helping to reduce information overload. Full-on machine reading based on unsupervised learning is now acknowledged as the new horizon, if Etzioni's heavily funded and highly anticipated MR research project KnowItAll is any indication. (Etzioni reports he cannot keep Google from hiring away his graduate students.) Such tools do much more than merely search and return information to users; they all identify novel relations and present then as hypotheses either implicitly or explicitly. Further, the tools lend themselves to rapid prototyping development. The most complex of such tools, Progol, is open source and freely available for commercial use; likewise, the other tools depend not so much on advanced code as they do on information science concepts, and so implementation of tools like these are reasonably attainable, particularly for multibillion-dollar organizations that depend on innovation.

Text mining promises to reduce information overload for pharmaceutical organizations by optimizing the way the organization finds novel ideas from the literature. Information overload within pharmaceutical organizations is driven by the mass quantity of data available to such companies. In addition to the research literature found in PubMed, other valuable knowledge sources exist, such as numerous searchable public genetic, genomic, proteomic, toxicological, chemical, epidemiological and pharmacological databases; internal high-throughput data stores and other research data; governmental regulatory records; data from clinical investigations and drug trials; international patent literature collections, and even internal corporate communications documents.

The problem of reducing information overload is in part an optimization problem because the goal is to derive the most valuable information possible from the sum total of all the knowledge sources available. It is not enough for any user of a system to just derive new information, much less re-find relevant but old information. The newly formed information must have the highest quality or utility possible.

If the information is to have the highest quality possible for drug discovery, it will need to possess at least some of the following features:

(a) reduce the amount of information needed for drug discovery research tasks;

(b) be generated relatively quickly;

(c) be based on as much information as possible namely through integration of large and disparate data stores;

(d) not overlap or violate property rights of other organizations;

(e) reuse old clinical data (originally gathered at incredible expense but highly valuable as it is human-, disease-, and compound-specific)

(f) be returned as input for further learning;

(g) lend itself to evaluation;

(h) be disease-relevant and highly-specific;

(i) involve currently tractable genetic, genomic, or proteomic mechanisms; and

(j) be entirely novel and possibly even contrary to common belief.

While the above may not be a complete list of attributes necessary for assessing the quality of information produced by a text mining system tuned to PGx drug discovery tasks, it seems to be a useable checklist nonetheless. If quality evaluation is not given priority, such systems pose the risk of increasing rather than reducing information overload. In this respect, the example of aardvarks and coffee is instructive. It is easy to use such a tool to create new information, but note that it is easy to generate information that appears to be relatively useless even if it seems informative.

The specific type of hypothesis we would want a text mining system to generate for PGx drug discovery would focus on the identification of relations between compounds, potential targets, and disease. Special care could be taken to identify compounds that have been successfully used for other ailments in order to see if they can be applied for new treatments. Text categorization tools could be utilized to filter items for a knowledge base as well as to identify features useful for a knowledge base. Clustering tools could be utilized to characterize the filtered contents. Finally, relationship derivation tools could assemble the best features for the most optimal use of all information sources available.

VI. Technology Adoption: Venkatesh's Unified Theory of Acceptance and Use of Technology

A. Background

One way of understanding how a company is using a technology is to devise a descriptive framework that helps describe the various features and potentialities of that technology. In the previous chapter's discussion of text mining, I formulated and described a model for understanding and interpreting text mining technologies as applied to PGx-based drug discovery problems. While such an approach helps frame any evaluation of text mining technologies, it fails to take into consideration the human dimension—uses, attitudes, and perceptions. Just as we may want to be able to describe the features and merits of a technology that a company is using, we may want to understand their technology adoption decision.

The Unified Theory of Acceptance and Use of Technology (UTAUT) (V. Venkatesh et al., 2003) is an attempt to model the use and desire to use new information technologies. Built from an analysis of other models of innovation and adoption[3], UTAUT accounts for 70% of the variance in user intent, a considerable improvement over the models from which it was built.

B. Model

The model claims that technologies are adopted for two primary reasons. One, because users intend to use the technology (*behavioral intention*), and, two, the users possess the means to use the technology (*facilitating conditions*). Three factors in turn determine intent: expectations about the system's performance (*performance expectancy*), expectations about the amount of effort required to use the system (*effort expectancy*), and outside human influences (*social influence*). The *age, gender, experience,* and willingness (*voluntariness of use*) of the individuals modify the effect of the four aforementioned factors. See Figure 2 below for an illustration of Venkatesh's UTAUT model.

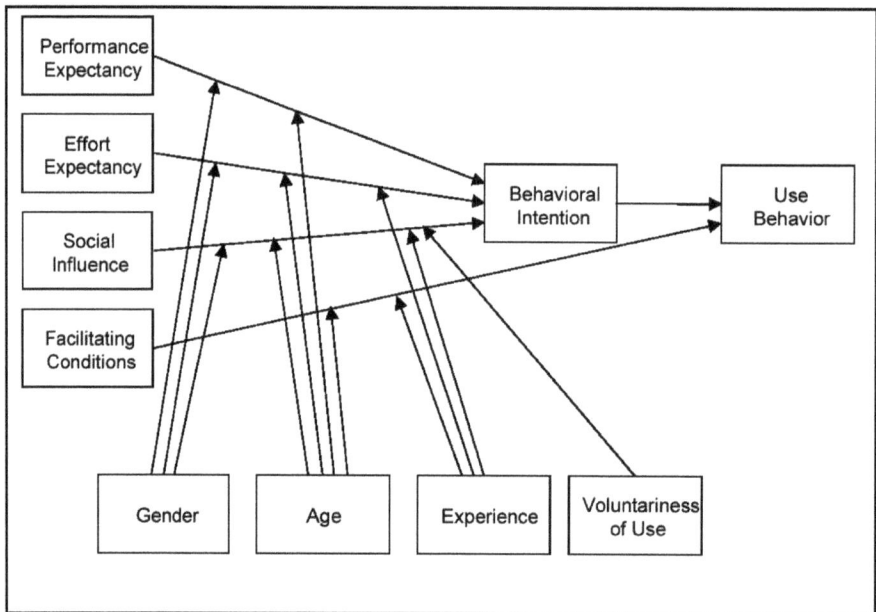

Figure 2 Venkatesh's UTAUT Model (V. Venkatesh et al., 2003)

C. Relevance to present study

The UTAUT model will be used in the present study in order to help frame a series of informal interview questions for PGx professionals responsible for making adoption-related decisions, particularly the adoption of text mining technologies for PGx-DD. Questions based on the UTAUT model will complement questions regarding specific details of current text mining technologies adopted (if any). Simply put, it is hoped that I can characterize not only the facts of any adopted systems but also any facts about the adopters and the context in which they work. The model is valuable because of both its relative simplicity and its predictive power. The UTAUT model facilitates the formulation of straightforward questions that should be relatively easy to evaluate.

VII.   A Brief Case Study of Text Mining Adoption for Pharmacogenomics-based
        Drug Discovery in a Large Pharmaceutical Company

    A. Case study design

        1. Main purpose

The main purpose of the case study is to evaluate the adoption of text mining to PGx-drug discovery problems in a large leading pharmaceutical corporation. Two topics comprise the focus of the study: the specific details of the organization's adoption of text mining, and the use of Venkatesh's UTAUT model to explain the organization's adoption. It is hoped that successes of and opportunities for text mining adoption in drug discovery may be illuminated while at the same time examining the UTAUT model.

        2. Design details

            a. Summary

Brief informal interviews of two leading adoption decision-makers in a large pharmaceutical corporation were performed in order to elicit details about their adoption of text mining.

            b. Interviewees

Two high-ranking professionals from one large pharmaceutical corporation agreed to interviews on the topic of text mining. One of the professionals[4] (*Researcher B*) is a leading innovator in PGx-based research, with a rich background in both clinical medicine and genomics research, having had great success at making breakthroughs with PGx-based target identification. The other professional interviewee (*Researcher A*) leads the team that implemented the company's first automated patent literature extraction system. Researcher A spearheads the organization's efforts to adopt advanced literature-based information technologies. Researcher A comes from decades of experience both in the research lab and in competitive intelligence.

c. Interview topics

Questions for the two interviewees focused on the following topics:

(a) their personal backgrounds and roles in the organization;

(b) their contributions to and use of text mining;

(c) details about their information needs and how they currently work to resolve them, with a particular focus on text mining and text mining-related technologies;

(d) their perceived needs for novel information and information overload-reducing technologies;

(e) personal and organizational attitudes towards text mining; and

(f) the general outlook for the future of text mining in drug discovery.

Two frameworks guided the interviews: the distinctions regarding PGx-DD and text mining provided above, and Venkatesh's UTAUT model. Ultimately, I wished to attempt to answer the following specific questions. What did the interviewees mean when they referred to something as "text mining"? How do their applications help them generate high-quality novel hypotheses? Finally, does the UTAUT model help explain their level of adoption?

d. Interview method

Each researcher was interviewed for 90 minutes via telephone. Before the interview, I had constructed a list of topic areas and possible questions. (Please see the appendix for the list of questions.) Topics and questions were often articulated or altered at the time of the interview in order to preserve the fluidity of the interview. I noted their responses in plain text files and sent them via email to each interviewee. Each interviewee was given the opportunity to review and alter their responses for accuracy and for confidentiality. The interviews were performed in August 2006.

B. Interview results

1. Background details of interviewees

Researcher A has worked in various research, development, and competitive intelligence roles in the pharmaceutical industry for over two decades. Researcher A received undergraduate degrees in biochemistry & molecular chemistry and worked in research labs at a small research institute for over a decade. Following work with the small research organization, Researcher A went to work for the large pharmaceutical company as a competitive intelligence (CI) analyst. Researcher A made the change from CI into a leadership role directing the use of advanced text information technologies for the pharmaceutical firm four years ago and has remained in that position ever since. Researcher A describes himself as a "middle manager" (Researcher A, 2006) as he reports to senior corporate officials with interest only in the results of the text processing technologies rather than the technical details.

Researcher B describes her role with the pharmaceutical company as that of an internal consultant. Researcher B works closely with people directing cardiopulmonary pharmacogenomics studies; it is her role to guide those research efforts while integrating human clinical research data with cardiopulmonary PGx data. Researcher B is an innovator in pharmacogenomics research. She has an MD/PhD from a prominent American university, having studied the genetics of heart disease under a renowned genetics/genomics pioneer. After finishing postdoctoral studies, Researcher B joined the large pharmaceutical company where she has been for the last several years. While Researcher B explains that her role in adopting text mining does not involve "buying robots" (Researcher B, 2006), her role involves adopting technologies and techniques to manage and analyze massive data sets that often include text. Researcher B has developed analytical techniques for the reuse of clinical data and its subsequent integration with PGx data.

   2. Mining technologies: use and factors affecting use

       a. Use and expectations about current use

According to Researcher A, the pharmaceutical organization has maintained interest in text mining-related technologies for at least seven years. Researcher A began evaluating text extraction technologies for pharmaceutical drug discovery while working in CI; he ran pilot studies to evaluate the technology's feasibility for drug discovery. The first text mining-related technology was launched in February 2005.

"As of [this year] we have about two years of tangible results" (Researcher A, 2006). Those positive results of using text mining-related technologies include the discovery of 20 new targets for the pharmaceutical pipeline along with the identification of *alternative indications* for compounds already used on known targets. (An *alternative indication* means that a given compound, discovered for other purposes, may be useful for a treatment different from the original).

The organization's own measure for successful adoption is simply the number of targets moved into high throughput screening. A set of 20 new targets for further PGx-based screening is considered a success by the organization as it will continue to expand its support of text-based technologies indefinitely. The effort, according to Researcher A, has been examined and praised by the company's R&D chief.

Information overload-reducing technologies are used regularly at the large pharmaceutical company. Researcher A's team has focused on using information extraction techniques (what he refers to as "text mining") for extracting valuable information from the international patent literature searching for alternative indications. Mining the patent literature poses an information overload challenge, according to Researcher A. "Our collections [of patents] run from 1000 to 55,000 patents" (Researcher

A, 2006). Patents are difficult to read as they are written "cryptically" (Researcher A, 2006). Rather than assign people to read through thousands of patents on an annual basis, Researcher A's team evaluates a small number of useful patents and then develops extraction templates that specify specific named entities or relational phrases in those useful patents. The templates are then utilized in an information extraction engine to identify matches across the multiple sets of patent literature collections. A subset of patents that match is manually evaluated for additional key phrase structures in order to broaden the match templates. This approach has allowed Researcher A's team to evaluate 500,000 patents in 18 months, a task that Researcher A estimates would take 50 man-years at the barest minimum if done manually. The technology allows the organization to keep pace with the growth of the international patent literature.

As stated in Chapter II, GSK exec Alan Roses described a problem with PGx-based discovery in that getting from 1) identifying all genes to 2) disease-specific disease-relevant targets was a huge gap, larger than initially understood. Each interviewee was asked for their opinion on the best way to close that gap.

Researcher A reported that he believes text mining is the means for closing the gap between the identification of all genes to the identification of disease-specific, disease-relevant targets. As Researcher A sees it, genes are targets. This simple conflation facilitates Researcher A's efforts to put together what he terms an *information package* for a specific disease. An information packet is a rich condensed bundle of information supporting the candidacy of a target for further evaluation. That package may contain statements from peer-reviewed literature, statistics, measures from the company's internal clinical data, and information from the patent literature.

Researcher A gives macular degeneration as an example of a disease. What his group will do to reduce information overload on macular degeneration (MD) is to locate the few known genes/targets related to MD; process all patents since 1986 and extract anything related to MD or the genes related to MD; and then use that information and evaluate it in light of the company's massive collection of clinical data. What is produced from each step is bundled together and the resulting information package is passed along for further review and PGx-related testing.

Researcher A reports both high precision and high recall in the use of the patent extraction system. In particular the >90% precision is considered a great success as any failed targets down the evaluative pipeline costs the organization greater and greater amounts of money. The process of evolving templates to get high precision is a costly and tedious task, Researcher A reports. Researcher A believes that sometimes the templates are too specific and too easily miss potentially useful information. Despite the shortcomings, Researcher A believes the benefits of the information extraction approach far outweigh the costs.

Researcher B routinely works with genetic association data for large populations, typically 1000-2000 people each. These clinical data sets contain genomic and phenotypic information about each person along with disease classes (whether they have or do not have a specific disease). Researcher B is working towards performing a metaanalysis by incorporating many clinical population studies together. Researcher B makes heavy use of logistic and linear regression analysis but has found that she may need a more highly dimensional approach; she is currently investigating incorporating more advanced pattern analysis techniques such as data and text mining into her efforts as a result.

Researcher B has utilized Researcher A's findings in the past and looks to continue to use them in the future. However, Researcher B mentioned that she feels that text mining in particular might have a difficult time in helping to locate novel targets. In particular, Researcher B is concerned with the problem of finding targets that are not in the research or patent literature. Researcher B feels that targets with a literature trail are unlikely to be novel, interesting, and hence useful targets for drug discovery. Researcher B also stated, "Most compounds go unpublished in the patent literature." (Researcher B, 2006) It is unclear to Researcher B how text mining might help identify novel target-disease relations from the research or patent literatures.

Both Researcher A and B agree that text mining the literature may be useful for identifying alternative indications. Both researchers described in some detail how alternative indications could be mined from the literature. The idea is to locate patented compounds in the patent literature, note the disease and targets specified in the patent, and then look for other targets and other diseases in the literature that are not specified in any patent. Both claim that the advantage of using alternative indications is that compounds that are already patented are likely to be already approved for human use and thus have a high probability of passing toxicity & safety screening again. (Patented compounds can be slightly modified to minimize toxicity changes while ensuring the patent has not been violated.) Researcher B added that if a compound that failed testing due to a lack of efficacy for a previous indication may also serve as a good candidate to revisit. The simple reason why alternative indications are promising is that when in the process of evaluating a compound for a specific use most if not all other specific uses are ignored entirely.

Integrating clinical data with patent and research literature is crucial for identifying compounds with alternative indications. At the time of the interview, Researcher B was evaluating ways to put together data from a large clinical study with patent and research literature for exploring alternative indications. Researcher B is currently leading a study of 6000 patients; each patient is being measured for as many phenotypes as possible. Researcher B noted that the reuse of older clinical data sets is limited by the smaller scale of phenotypic assessment, as fewer phenotypes have been recorded in past studies.

Researcher B noted that another crucial feature of targets, after novelty, is that of tractability. "From the genome project we know 30,000 genes, but which of those can be targets?" (Researcher B, 2006). She noted that tractable targets tend to code for enzymatic activity, cell signaling receptors, or ion channels, thus reducing the set of 30,000 genes considerably to around 3000-4000. Researcher B speculated that text mining may be able to be used to help "identify the most tractable element of a disease process" (Researcher B, 2006). Researcher B refers to the tractable genome as "low hanging fruit" (Researcher B, 2006).

Researcher B expects that the clinical data she is now collecting, along with old clinical data repositories, will be useful for drug discovery for years. "We'll probably be analyzing this data for a generation," Researcher B remarked (Researcher B, 2006). While the data in the short run may lead to determining the role of individual phenotypes in disease, the much more difficult part is to develop an understanding of the interactions. Researcher B feels strongly that pattern learning techniques may be the only way to understand how combinations of phenotypes may better explain diseases. However, Researcher B feels that the barrier to evaluating interactions for as many as 500,000 dimensions is computational power. She feels that the limits of processing power may be a primary reason for why the data may continue to be evaluated for the next few decades.

b. Factors influencing further adoption

i. Age, gender, experience and voluntariness of involved decision-makers

When asked about the age and relative experience of other decision makers involved in text mining adoption, both respondents stated that age and experience varied widely with an average of approximately 10 years in the organization. Researcher A commented that most of those participating in adoption decisions are at a director level or higher. Researcher B noted that the senior decision makers tended to be about 75% male to 25% female while those involved in genetics were a little closer to 50/50 male/female.

72

Voluntariness of use refers to the willingness of a potential user to use a particular system. Both respondents enthusiastically reported that most of the people involved in adoption decision-making are highly motivated people who are doing what they want to do. Both A and B felt strongly that most of those involved have firm intrinsic and extrinsic motivating factors that includes significant inner drive for accomplishment, professional recognition, and monetary reward.

Both Researcher A and B agreed that they find themselves in command of their adoption decisions. Researcher A commented, "nothing's being force-fed to us" (Researcher A, 2006). The voluntariness of participating in the adoption is high according to both interviewees. Researcher B added, "keeping your job is good but pioneering genetic breakthroughs is a huge motivator" (Researcher B, 2006). She added that the potential for discovery making is tremendous given the large scale of data becoming available in conjunction with the recent development of techniques such as text mining that are poised to leverage that scale effectively.

ii. Performance expectations

When asked about expectations about their patent extraction system adoption, Researcher A answered that the system simply was expected to outperform manual patent analysis. Responded A added that those expectations have been exceeded to the point that his group has begun to develop a more robust information extraction system that uses MEDLINE as its input. The system was designed to influence key decision points in drug discovery, says Researcher A, and he says the current system has met that design requirement.

Researcher B reflected more on the possibility of adopting true text mining and pattern learning technologies when asked about performance expectations. Researcher B is currently evaluating support vector machines and supervised learning techniques for use in managing the massive amounts of data she faces daily. Her expectations of the performance of both data and text mining are very high. The reasons Researcher B cites in order to justify her high expectations include gains in efficiency; more precise and relevant results; using, reusing, and integrating multiple data stores; and, with particular respect to text mining, the discovery of new uses for the data. Researcher B reports that it is unclear to her how mining data and text can provide novel information, a concern that lowers her expectations to some degree.

iii. Expectations about level of effort

Researcher A reported that everyone involved with the adoption of the patent mining tool in 2005 knew the task "would be complex" (Researcher A, 2006). However, the group mitigated the adoption with a hybrid build/buy approach to the software. His group purchased commercial extraction software and then heavily customized the application for their specific needs. The software they purchased, Researcher A said, incorporates user-friendly interfaces that made learning the application easier. When the group did their due diligence on commercial information extraction software, the group made user-friendly interfaces a top priority specifically to ease adoption and use. Researcher A added that the complexity was not a deterrent, as none of the individuals involved in adoption reported that they were slowed by complexity of any sort. Given that the system's output is simple, complexity has no impact on key decision makers who have influence on future adoption decisions. If anything, the simplicity of the results has served as a support for future adoption. Researcher B commented that her only concerns with complexity regard the time required to design, develop, and test advanced text mining applications. "Time is a bottleneck," Researcher B added (2006).

iv. Social influence

Researcher A reports that *social influence* played a role in his choice to adopt the information extraction technology. "10,000 patents in a week was particularly impressive [to others] in light of what people could do manually" (Researcher A, 2006). The system continues to impress, as it has influenced other groups in the company, such as a safety assessment group, to adopt information extraction technologies.

Researcher B has a different take on social influence, particularly with senior decision makers. Researcher B notes that most of the time senior officials do not know much about the system, are not much interested in the details, and are not interested in the creativity involved. Instead, they are focused on the results and are not so much as impressed as they are satisfied.

v. Facilitating conditions

Both Researcher A and B report that they have more than sufficient resources for adopting text mining adoption. Researcher A explained that both the organizational structure as well as the material resources are more than adequate for successful adoption and impose no barriers. Researcher A reports that in particular the computer hardware invested in the patent mining project alone exceeds $1million US.

Researcher B, however, reportedly feels more in need of labor for any successful mining adoption. Further, "higher-ups do not understand the scope of the problems," Researcher B added. The organization has made a shift away from doing their own target identification and is beginning to outsource such early discovery steps. Instead, the organization is shifting its focus and labor on to Phase II and III pipeline projects. Researcher B feels that discovery itself is being neglected.

VIII.   An Analysis of Case Study Interviews

A. Introduction: interviewees and aims

Through interviews of pharmaceutical professionals, I hoped to utilize my own text

mining framework as well as Venkatesh's UTAUT model in order to understand and

evaluate text mining adoption and innovation in a real-world setting. The two

interviewees provided two complimentary ways of looking at text mining adoption for a

pharmaceutical firm: a PGx-oriented scientist innovating new analytical techniques while

making discoveries and an individual responsible for deploying information overload-

reducing technologies for the purposes of making discoveries from large text-based data

collections. Both interviewees proved to be good subjects for the present study.

I designed the interviews to operate on several levels. The first level of operation was for

gaining a basic understanding of the organization's text mining adoption. I presumed that

the interviews might help identify the specific problems their users are trying to solve, the

relevant problems the interviewees are anticipating, and the intentions motivating the

efforts to solve present and future problems. On the second level, I hoped that the

organization properly understood as a market leader would provide an example of state-

of-the-art information-based discovery in the pharmaceutical industry. Finally, on the

third level, I wanted to evaluate the UTAUT model when applied to a real-world case of adoption.

B. Current and expected text mining use

In Chapter V, I distinguished text mining from information extraction, information retrieval, and data mining. Most controversial was distinguishing text mining from information extraction. Many people consider information extraction to be a form of text mining. I wished to differentiate IE from text mining in order to foreground a critical distinction: *generating novel information* versus *extracting information that already exists.* Text mining generates new information using current information as an ingredient while information extraction merely finds existing information.

The distinction I make between text mining and information extraction is slight in some respects yet it is pivotal. Given the present study's definition of text mining, the extraction of words and other linguistic features is reduced to an intermediary step on the path to the creation (generation, synthesis, derivation) of new information. In essence, by redefining text mining and distinguishing it from IE I have attempted to reframe any discussion of text mining and other information overload-reducing technologies. Once a set of external domain rules of any kind (whether ILP or machine learning algorithms) is applied to extracted text nuggets for their reassembly the task of finding is eclipsed by the act of creation. The whole is more than the sum of its parts. In turn, the act of creating new information raises the importance of the user's role in specifying problems and potentially satisfactory solutions. More importantly perhaps is realizing that the difference between text mining as it is presently defined and IE is precisely the difference between innovation and adoption itself: the creation of new ideas versus finding and leveraging already-established ones.

From the interviews, we learned that the organization might not have adopted text mining technology in the sense of text mining presented here. Rather what has been implemented by the organization is information extraction. To put it in another way, the organization may not be an innovator of text mining solutions.

The difference between thinking someone has adopted text mining when in actuality adopting only IE and actually adopting text mining at its current capacity is a *conceptual* difference. The conceptualization of information extraction as text mining leads people to believe they have actually adopted text mining when, according to the definition of text mining provided in the present study, they have merely adopted information extraction. In common parlance, the organization has adopted text mining. Yet when the concept of *generating novel information* is brought to the fore of text mining's meaning, it becomes more difficult to admit they may have adopted text mining. It is not entirely clear that the organization's IE tools provide the organization with the best means to generate (rather than merely find and extract) the most optimal information given their needs. The apparent lack of conceptualizing text mining as an innovative means of automatically generating novel information is a leading barrier to actually innovating text mining solutions.

Conceptualizing IE as text mining also creates a practical problem. Information extraction has the virtue of high precision and a high measured recall yet, in reality, it will have very low recall since the true semantic matches will have a far broader set of patterns than can be manually identified. Researcher A reported considerable manual efforts to detect and configure extraction patterns (2006). The manual task of identifying positive linguistic patterns appears to emerge as the focus and the measure for success. Less time as a result will be spent on further efforts to evaluate the extracted information on the basis of specific quality standards such as the ones specified in section E of Chapter V, namely since manual effort—human learning rather than machine learning-- will be more focused on pattern learning tasks.

By adopting an IE system, however, the organization has adopted core text mining technologies to some extent. The system put in place by the study subjects does process large literature collections. It extracts important information that is highly specific and disease-centric. It augments the manual assembly of novel relations. The organization has in place an advanced information extraction system just shy of what is presently defined as text mining. The organization has much of the expertise in place as well as hardware sufficient for upgrading to text mining. The differences between the system adopted by the organization and a text mining system as defined in the present study rest with the automation of pattern learning, relationship construction (specifically relationships between isolated pieces of text information from different documents), and information quality evaluation. If the three features were present, their combination would free up more time and energy to allow for further manual exploration of novel information for PGx-DD. Researchers such as Researcher B would be enabled to spend more time on doing what they do best.

Researcher B reported her concern about text mining's ability to produce truly novel information. I can only speculate on the following point, but it may be that Researcher B's concerns about novelty arise from the limitations of their current system to generate novel statements. While the system used by her organization has generated targets for further evaluation, it has done so at the rate of approximately one per month. The rate of the current information extraction system may outperform more traditional and manual approaches employed by Researcher B such as tedious regression analyses of thousands of variables or manual research literature review. However, that rate may not be optimal given business needs. If we estimate expenditures to date for Researcher A's system at approximately $2 million for hardware, software, and labor, each target identified cost approximately $100,000. Researcher B reports similar productivity via her more manual methods. Neither approach seems to be outperforming the other. However, contrasting the two approaches seems the wrong way to frame the discussion. Tools that augment, accelerate, and complement Researcher B's methods rather than compete with them likely comprise the better means for accelerating drug discovery. A text mining system— one that meets the present study's definition of text mining as that which produces novel hypotheses for a scientist's further examination—would be precisely the sort of system that might best augment Researcher B's work.

Increasing the number, size, and diversity of inputs for text mining adoption is one of the major text mining development trends indicated by both interviewees. Researcher A referred to his efforts to expand IE beyond the patent literature into the medical research literature. Researcher B mentioned that new clinical research data is coming in at greater and greater scales while she is repurposing older clinical research data to new studies. Integrating multiple inputs may play an increasingly greater role in the application of text mining to drug discovery.

Everett Rogers' *Diffusion of Innovations* (Rogers, 2003), first published in 1962, helped form the intellectual basis of Venkatesh's UTAUT model (V. Venkatesh et al., 2003). While Venkatesh's model focuses on technological adoption, particularly IT adoption, Rogers' innovation model centers on the distribution of technological adoption over time. Rogers differentiated people by the timing of their adoption of a specific technology relative to time at which the technology was innovated. The earliest of adopters Rogers labeled *innovators*. Later adopters, in order, include groups Rogers termed as *early adopters*, *early majority*, *late majority*, and *laggards* (Rogers, 2003).

Rogers defines *innovation* both as a type of adoption and as an act of creation preceding adoption. Regardless of whether *innovation* and *adoption* are disjoint they are at least conceptually intertwined for Rogers. Rogers defines innovation as, "an idea, practice, or object" that is new to a "unit of adoption" where a unit of adoption is either an individual or group with a similar purpose (Rogers, 2003, p. 12). *Innovativeness* as Rogers defines it is, "the degree to which [a] [...] unit of adoption is relatively earlier in adopting new ideas than other members of a system" (Rogers, 2003, p. 267).

Rogers writes that the possession of an innovation, that is, *a novel concept*, creates

doubts. "Will the innovation solve an individual's perceived problem?" (Rogers, 2003, p.

14) Information-seeking in order to mitigate the doubts about an innovation create a new

risk of falling from innovation into early adoption. In many cases, innovation necessitates

a high level of risk. Rogers characterizes the innovator as essentially *venturesome,*

possessing "a desire for the rash, the daring, and the risky" (2003, p. 283). The innovator

"plays a gatekeeping role in the flow of new information into a system" (Rogers, 2003, p.

283).

While innovators launch new ideas, *early adopters* serve as opinion leaders for their

adoption units. They serve as role models for later adopters and help trigger popular

acceptance. Early adopters are in a sense leaders in conventional wisdom and display

their approval through adoption. Where innovators are less esteemed in their locales,

early adopters are the most esteemed (Rogers, 2003).

By Rogers' standards it appears that Researcher A and B as well as their organization are

not exactly *innovators* but rather *early adopters* of text mining applied to PGx-DD. While

both researchers' responses indicate an ability to understand and apply complex technical

knowledge, their responses also indicate a slight unwillingness to make venturesome

technological risks that have a high likelihood of generating losses.

The organization's adoption of IE, a technology that has been in existence for decades,

followed a significant amount of deliberation and exploration (Researcher A, 2006).

Further, both the interviewees and their organization seem to function in a role of opinion leadership and social respect, and both of the interviewees pay attention to success. They epitomize the very definition of Rogers' *early adopter*. Researcher A likely knows the technological steps necessary for innovating yet he appears to be operating within the parameters of the organization that seem to dictate early adoption rather than the innovation. Likewise, Researcher B's ambitions with text mining are likely conscribed by an institutional lack of willingness to support innovation.

C. From case to industry: PGx-specific problems

From the interviews, it appears that the present study began with both a good inventory of problems best solved with text mining and a good sampling of working solutions. Issues of alternative indications, data reuse, disease specificity, and novelty repeatedly arose in the interview conversations. They confirm the industry-wide problems best addressed using text mining that were first referred to in section E of chapter V. The interviews also corroborate the notion that the organization suffers from the same sorts of information-centric PGx-DD problems as the rest of the industry.

Uncovering alternative indications appears to be a particularly rich area for text mining in the pharmaceutical industry. Both interviewees demonstrated marked enthusiasm about the prospect of mining alternative indications. Alternative indications make for inexpensive drug development cycles. Fewer new studies need to be run before approval; compounds that show alternative indications previously have been shown to be safe and efficacious for treatment of human disease. Efficacy and toxicity are the two major reasons drugs fail to reach market. The solution to programmatically uncovering alternative indications rests with automatic generation of novel hypotheses given data inputs from sources as diverse as the patent and research literatures as well as genomics, proteomics, chemical, toxicological, pharmacological, regulatory, and clinical databases.

Novelty remains an elusive concept for text mining adopters in PGx-DD. While Researcher A placed little emphasis on the role of novelty in evaluating the output of an IE system, Researcher B voiced doubts as to whether an information tool can in effect find what is not there. Performing extraction tasks on the patent literature is a necessary step in ensuring the lack of something in the patent literature. However, it is unlikely to serve as a sufficient step. Novelty remains elusive in part because the other necessary pieces remain largely under-explored. The other necessary pieces are likely to include general linguistic features as well as context-specific features yet they remain unidentified.

D. Applicability of UTAUT to innovation

In a casual conversation with a friend the other day, I explained the UTAUT model in simple terms. UTAUT is a model that predicts the use of a technology on two main factors: the intent to use the technology along with the facilitation of doing so. Therefore, according to the UTAUT model, if you want to start using Google, then if you have not only the time to try but also the necessary equipment, you are going to start. My friend immediately questioned the need for such an obvious theory, saying it was so obvious that it really did not require articulation. What I then described was that which makes the

UTAUT theory interesting: the three factors that indicate intent and the factors that influence those features of intent.

The UTAUT model states in short that expectations about performance, expectations about effort, and social influence heavily determine the intent to adopt a technology. The model also states that a person's gender and age heavily influence all three factors, that experience influences effort expectations and social influence, and that voluntariness heavily modifies social influence. The interviews demonstrated that at least according to UTAUT all conditions are in place for adoption of text mining in the company of interest. Despite the presence of all necessary and sufficient conditions, the organization has not fully adopted text mining, not at least in the sense of text mining defined in the present study.

Both Researcher A and B reported very high performance expectations for text mining for drug discovery. They both reported that concerns about effort were not deterrents either. Social influence factors seemed to encourage success, particularly when we examine their voluntariness. Researchers A and B both described their company as an organization that gives wide latitude to self-initiation to the point that the voluntariness is expected. Researchers A and B are expected by their employers to devise solutions independently. Both reported that the terms of their solutions are rarely if ever forced upon them.

The intent to use text mining is clearly in place. The facilitating conditions that permit actualization of the intent are also present. Researcher B has reported having no difficulties in securing the optimal amount of computing hardware, software, and labor to succeed at previous text mining-related tasks. Both describe an organization willing to provide sufficient monetary and organizational support for new technologies that promise gains in efficiency. All of these factors for adoption are in place, and there is to some extent a degree of adoption already. Yet text mining as defined in the present study has not been adopted to its fullest. The UTAUT model does not explain how text mining has not been adopted by the organization other than by simple statistical variance.

The UTAUT model of adoption was selected over Everett Rogers' innovation model because UTAUT was tuned to the peculiarities of information technology, while Rogers' theory was a more general one, derived in large part from case studies of farmers innovating and adopting new agricultural techniques. While I cannot at this point state that Rogers' model would have been better for the present study, Rogers' notion of innovation as distinct from adoption (see section B above for an earlier discussion) helps explain what is missing from the UTAUT model given the present case study. While every element of the UTAUT model was in place at the pharmaceutical organization to suggest adoption, it appears that the organization's *conceptualization* of text mining undermines its fullest and most timely adoption. Seeing text mining as equivalent to information extraction when adopting information extraction leads the organization to believe it has adopted text mining. It also encourages the organization to neglect text mining's most powerful feature: the automated generation of high-quality novel information. Because the organization does not understand text mining as an automated

87

means for generating novel ideas, the organization cannot adopt core text mining functionalities, much less innovate entire text mining solutions.

In order to be adopters of text mining as defined in Chapter V, the firm would need to be able to support innovation. The firm supports adoption, at least by the virtue of the fact that it thrives on the innovation of new drug treatments for human disease. The organization, however, does not appear to possess the institutional will or knowledge necessary to innovate text mining solutions for PGx-DD. The application of the UTAUT model to the present study highlights insufficiencies in both the UTAUT model and in the company itself: innovation. It is only with Rogers' work on innovation we can begin to recognize UTAUT's shortcoming, and, in turn, the pharmaceutical company's shortcoming.

The finding of a lack of innovation in both the firm and the UTAUT model is nevertheless inconclusive. I based the case study on interviews of only two professionals from a large company that has thousands of employees. Further, the interviews were brief. Additional interviews could uncover information that invalidates any conclusions I may have reached regarding any institutional lack of innovation. The firm may in fact be innovating with text mining applied to PGx-based drug discoveries. Simply stated, I do not possess sufficient knowledge of their proprietary activities in the text mining domain to feel secure in my conclusion. Likewise, the UTAUT model states that it captures only 70% of the variance of adoption. It makes no claims with respect to innovation over adoption and similarly makes no claims to completeness. The present study does not bring into question the UTAUT model. Rather it only brings into question its applicability to one specific case of innovation.

IX. Conclusions

Innovation is a core element of the pharmaceutical industry, and drug discovery is perhaps the quintessence of that innovation. Many in the pharmaceutical industry have hoped for a long time that pharmacogenomics would bring dramatic breakthroughs in drug discovery. Pharmacogenomics has instead brought information overload.

Text mining is uniquely positioned to reduce information overload and help solve core PGx-DD problems. In particular, text mining can help with the discovery and identification of highly specific medical needs, the identification of tractable drug targets, and the discovery of NMEs. It can even help screen lead compounds for patent infringement, toxicity, and efficacy. In some areas, the problems facing PGx-DD are scientific ones, such as the need for more functional analysis data. However, other problems arise from information management and processing needs, needs that text mining can help meet. More importantly, text mining at its most advanced begins to break down the divide between science problems and information problems, as Muggleton's robot scientist illustrates. It appears the lines between lab work, data evaluation, hypothesis formation, study design, and text mining have forever been blurred.

Text mining is certainly not new to the pharmaceutical industry, as the case study illustrates. Nevertheless, I found only information extraction where I expected to find text mining. I now know that the goal of text mining is different from information extraction: *to learn something new*, rather than *to find something neglected*. I have learned that text mining relies on pattern learning rather than merely pattern extraction. Finally, I have discovered that text mining automatically generates novel information that in turn can be evaluated based on quality standards. Text mining can perform more discovery subtasks than information extraction. Text mining is better equipped to help accelerate critical thinking tasks. I have seen in Muggleton's ILP work, in Swanson & Smalheiser's Arrowsmith, and in PubMiner, some concrete examples of high quality information generation and research support tools. I did not find such innovative use of text mining in the case subject, however. I found a case of adoptive use, yes, but not innovation.

I learned from our interviewees that the best quality information can be generated only by including as many disparate information stores as possible. Both interview subjects also cited the importance of information novelty relative to patent infringement. As a result, in the future I anticipate tools that can utilize old clinical data. Further, I expect a comprehensive text mining system that is able to generate new input for its own system, just like the robot scientist. I expect such software to arrive at new ideas given old facts and to prioritize these novel hypotheses according to their potential utility for drug discovery. Finally, I believe automated text mining will provide the best means to verify the novelty of our machine-generated scientific insights. Verifying the novelty of a scientific claim manually became an intractable problem many years ago. As competitors tune their information extraction tools to the task of automatically detecting patent infringements, the stakes for ensuring novelty rise ever higher.

From the case study, I learned that the difference between adoption and innovation resides not only in timing, personalities, and organizational context, but also in the ability to conceptualize a new solution. I have also discovered that conceptualization differentiates innovative text mining applications from its predecessors and cousins such as information extraction tools. Rapidly synthesizing new ideas out of the pieces of old ideas is perhaps the very essence of innovation. Leading pharmaceutical companies are organizations that are the leading innovators of drugs. They innovate drugs at a higher frequency and a higher quality: they bring more drugs to market per year, drugs that often meet the needs of large groups of people. Given the potential for rapidly accelerating drug discovery through text mining, merely adopting information extraction technology is equivalent to giving up altogether on the central mission of the pharmaceutical organization: to innovate. Outsourcing the earliest elements of the drug pipeline—outsourcing innovation itself—is equivalent perhaps to an industry betraying its own core competencies. Building innovative text mining systems that generate novel information is fast becoming equivalent in drug discovery to innovation itself.

We stand on a precipice of treating numerous diseases; we have billions of points of information all practically begging to be put together for the use of scientists creating new treatments. Rather than separating centers of information and knowledge, pharmaceutical companies must strive to bring together information at a scale and

92

dimension never seen before, expanding even to hospital-generated clinical data, rather than dividing such interests among smaller segregate companies. Pharmaceutical companies possess sufficient motivation and means for bringing data together to automatically generate innovations. The success of pharmaceutical organizations depends upon their ability to innovate with the very means of automated innovation itself. Ultimately, it seems, our own lives depend upon it.

A. Notes

[1] PCR - Polymerase Chain Reaction; a widely used genetics research technique that allows for the rapid synthesis of millions of copies of a DNA sequence of interest. PCR allows for easy identification of a specific sequence of interest within any biological sample as it allows its user to specify a complimentary base sequence at the outset. If the complement to the complimentary base sequence is present before PCR then the PCR, if properly performed, will produce an easily detectible high volume of the sequence of interest (referred to as *amplification*). If the sequence is not present, no amplification will take place; the technique was invented by Kary Mullis in 1983, earning him the 1993 Nobel Prize in Chemistry.

[2] Incidentally, the use of Arrowsmith in evaluating the relationship between coffee and aardvarks supports a number of novel hypotheses. Among them is the notion that proximity of coffee plantations to aardvark populations could promote the proliferation and spread of leishmania among humans; young phlebotomine sand flies can feed from the sugars of coffee beans until they are mature enough to begin feeding from the blood of aardvarks. For the purposes of human health, it may be important in places like Kenya where such a scenario is possible to keep aardvarks and other small mammals that do not eat coffee out of coffee farms. It also suggests that using insect-eating mammals to control pest infestation of coffee fields may spell an increase in leishmania among humans in the surrounding areas. At once, I marvel at the power of such a tool and doubt its ability to reduce information overload

[3] UTAUT is constructed from the Theory of Reasoned Action, the Theory of Planned Behavior, the Technology Acceptance Model, and Innovation Diffusion Theory.

[4] Numerous steps were taken to protect the confidentiality and privacy of the interviewees to minimize or eliminate any risks they may face by their participation. The case study was evaluated and approved by the University of North Carolina's Institutional Review Board (IRB) in March 2006. In the spirit of protecting the interviewees, they will simply be referred to as Researcher A and Researcher B. Researcher A will be referred to arbitrarily as a male and Researcher B will be referred to arbitrarily as a female.

B.   Appendix: Interview Outline and Candidate Questions for Informal Case
        Study Interviews

1. Summary

The purpose of this research study is to learn about the adoption of text mining
technologies for pharmacogenomics-based drug discovery efforts. The study will
comprise my Master's Paper in order to satisfy requirements for the Masters of Science in
Information Science at UNC-Chapel Hill's School of Information and Library Science.
Your participation is deeply appreciated.

2. Instructions

This is an informal and unstructured interview concerning the application of text mining
to pharmacogenomics-based drug discovery. Feel free to answer questions as you see fit.
If you feel any of this process may breach your confidentiality, put you at risk of psychic
or economic harm, or push you towards any legal jeopardy (particularly by encouraging
you to violate the proprietary nature of your employer's information), you are encouraged
to report this to me, ask to change the question, or even stop the interview if necessary.
Questions that may pose a risk to you, were you to answer them as asked, will be avoided
by the interviewer. You have the right to refuse to answer any question, and further, you
have every right to retract any answer should it pose any risk. In order to accomplish this

revision, you will be given a further opportunity to retract, modify, or add to comments in this interview through a written revision process we will conduct via private email immediately following the interview. There may be uncommon or previously unknown risks. You should report any problems to the researcher.

3. Interview Questions

a. General context details

- Please describe some general & non-identifying aspects of your educational and professional background and education; experience with pharmacogenomics; role in decision-making with respect to pharmacogenomics and/or information technology adoption.

- Your number of years of experience?

- Your role in making adoption decisions?

- Informatics & statistical analysis technologies adopted & currently utilized?

- How do you use text mining: inspirational idea-provocation or as something that provides a distinct line of evidence for a candidate target?

- Do you use statistical mining-based text analytics or information extraction-type applications?

- In what ways do you believe that text mining might reduce the problem of information explosion?

b. Venkatesh's Model: Predicting Use= intent + facilitating conditions (resources) modified by gender, age, experience, and voluntariness of use

      i.   On predicting intention (performance expectancy, effort expectancy, social influence)

- How useful do you think text mining is?

- How well does text mining fit your job or help you do your job?

- What is the advantage of text mining relative to its precursors?

- What are your expectations about using text mining?

- How difficult is the task of using text mining tools?

- Does the complexity of the text mining tools you use dissuade people from using the tools?

- Do you believe your current system is easy to use & make conclusions from?

      ii.   On facilitating conditions (resources)

- Is your organizational infrastructure sufficient or deficient for adoption? If so, how?

- Is your technical infrastructure adequate or otherwise? How?

      iii.   On modifiers of predictors

- What is the ratio of male/female among people making adoption decisions?

- What are the ages of others involved in adoption decision-making? Is there an average age or is it widely varied?

- On average, how many years of experience do the other decision-makers have? Range?

- Are the people working with text mining and making adoption decisions, are they the sort of people who want to be there, or do they suffer from a sort of 'day job' syndrome?

- The new text mining and text-related technologies you adopt, do they tend to be technologies you devise and establish, or do they tend to be rather decreed and passed down?

- Do your peers think you should use such technologies?

- Does using text mining look good to others? Is it impressive?

C. References

Accenture. (2003). *Pharmaceutical companies see balance of power shifting.* Retrieved

   March 10, 2006, from

   http://www.accenture.com/Countries/South_Africa/Research_and_Insights/Pharmac

   euticalShifting.htm

Agrawal, R., Imielinski, T., & Swami, A. (1993). Mining association rules between sets

   of items in large databases. *ACM SIGMOD International Conference on

   Management of Data,* 207-216.

Altman, R. B., & Klein, T. E. (2002). Challenges for biomedical informatics and

   pharmacogenomics. *Annual Review of Pharmacology and Toxicology, 42*(1), 113-

   133.

Bawden, D., Holtham, C., & Courtney, N. (1999). Perspectives on information overload.

   Aslib Proceedings, 51(8), 249-255.

Breckenridge, A. (1996). A clinical pharmacologist's view of drug toxicity. *British

   Journal of Clinical Pharmacology,* 42(11), 53-58.

BTG. (2006). *The innovation gap.* Retrieved March 28, 2006, from

   http://www.btgplc.com/compliant/btgsbusiness/theinnovationgap.cfm

Cardinal, L. B. (2001). Technological innovation in the pharmaceutical industry: The use

   of organizational control in managing research and development. *Organization*

   Science, 12(1), 05 March 2006-19-36.

Collica, R. S. (2003). Mining textual data for CRM applications. *DM Review*(July), 02

   October 2006 . Retrieved 02 October 2006, from

   http://www.dmreview.com/editorial/dmreview/print_action.cfm?articleId=7037

Crick, F. (1970). Central dogma of molecular biology. *Nature, 227*, 561-563.

Cunningham, M. J. (2000). Genomics and proteomics: The new millennium of drug

   discovery and development. *Journal of Pharmacological and Toxicological*

   Methods, 44(1), 291-300.

DiMasi, J. A. (1999). New drug development in the united states from 1963 to 1999.

   Clinical Pharmacology and Therapeutics, 69, 286-296.

Drews, J. (2000). Drug discovery: A historical perspective. *Science, 287*(5460), 1960-

   1964.

Ekins, S., Bugrim, A., Nikolsky, Y., & Nikolskaya, T. (2005). Systems biology:

   Applications in drug discovery. In S. C. Gad (Ed.), *Drug discovery handbook* (pp.

   123-183). Hoboken, New Jersey: Wiley Interscience.

Ellingsworth, M., & Sullivan, D. (2003). Text mining improves business intelligence and predictive modeling in insurance. *DM Review*(July), 02 October 2006 . Retrieved 02 October 2006, from

http://www.dmreview.com/editorial/dmreview/print_action.cfm?articleId=6995

Eom, J. H., & Zhang, B. T. (2004a). BioPubMiner: Machine learning component-based biomedical information analysis platform. *Intelligent Information Technology, Proceedings, 3356*, 11-20.

Eom, J. H., & Zhang, B. T. (2004b). PubMiner: Machine learning-based text mining system for biomedical information mining. *Artificial Intelligence: Methodology, Systems, and Applications, Proceedings, 3192*, 216-225.

Etzioni, O., Banko, M., & Cafarella, M. J. (2006). Machine reading. Paper presented at the *Proceedings of the 21th National Conference on Artificial Intelligence (AAAI 2006),* Boston. Retrieved 05 October 2006, from

http://www.cs.washington.edu/homes/etzioni/papers/aaai06.pdf

Fenstermacher, D. (2005). Introduction to bioinformatics. *Journal of the American Society for Information Science and Technology, 56*(5), 440-446.

Gad, S. C. (2005). Introduction: Drug discovery in the 21st century. In S. C. Gad (Ed.), *Drug discovery handbook* (pp. 1-10). Hoboken, New Jersey: Wiley Interscience.

Gibson, G., & Muse, S. V. (2004). *A primer of genome science.* Sunderland, Massachusetts: Sinauer Associates, Inc.

GlaxoSmithKline. (2006). *GSK 2005 annual review* (Annual ReviewGlaxoSmithKline.
   Retrieved March 10, 2006,

GlaxoSmithKline. (2005). *Annual review 2004* (Annual ReviewGlaxoSmithKline.

Goldman, R. J. (2003). Technology transfer in rehabilitation: A personal account. *Journal
   of Rehabilitation Research & Development, 40*(2), 28 March 2006

Hearst, M. (2004). Text data mining. In R. Mitkov (Ed.), *The oxford handbook of
   computational linguistics* (pp. 616-628). Oxford, UK: Oxford University Press.

Hearst, M. (2003). *What is text mining?* Retrieved 28 September 2006, 2006, from
   http://www.ischool.berkeley.edu/~hearst/text-mining.html

Herron, P. (2005). *Using WordNet in document clustering of a consumer health web
   collection.* Retrieved September 28, 2006, from
   http://www.unc.edu/~pod/papers/HerronP_WordNetBasedDocumentClusteringForC
   onsumerHealthPortal.pdf

King, R. D., Whelan, K. E., Jones, F. M., Reiser, P. G. K., Bryant, C. H., & Muggleton,
   S., et al. (2004). Functional genomic hypothesis generation and experimentation by a
   robot scientist. *Nature, 427*(6971), 247-252.

Lander, E. S., & Weinberg, R. A. (2000). GENOMICS: Journey to the center of biology.
   Science, 287(5459), 1777-1782.

Lichtenberg, F. R. (1998). *Pharmaceutical innovation, mortality reduction, and economic growth* (NBER Working Paper No. W6569). Cambridge, MA: National Bureau of Economic Research. Retrieved 01 March 2006,

Light, D. W., & Warburton, R. N. (2005). Setting the record straight in the reply by DiMasi, Hansen and Grabowski. *Journal of Health Economics, 24*(5), 1045-1048.

Luhn, H. P. (1958). The automatic creation of literature abstracts. *IBM Systems Journal, 2*(2), 159-165.

Mittermayer, M. (2004). Forecasting intraday stock price trends with text mining techniques. *Proceedings of the 37th Hawaii International Conference on System Sciences,* Hawaii. , *3*(3) 30064.2.

Muggleton, S. (2003). *Inductive logic programming.* Retrieved 10/04, 2006, from http://www.doc.ic.ac.uk/~shm/ilp.html

Muggleton, S. (1999). Scientific knowledge discovery using inductive logic programming. *Communications of the ACM, 42*(11), 42-46.

Muggleton, S. (1990). Inductive logic programming. *New Generation Computing, 8*(4), 295-318.

National Institute for Health Care Management Foundation. (2002). *Changing patterns of pharmaceutical innovation.* Retrieved 01 March, 2006, from http://www.euractiv.com/Article?tcmuri=tcm:29-117668-16&type=Analysis

Ng, R. (2004). *Drugs :From discovery to approval.* Hoboken, N.J.: Wiley-Liss.

Oxagen Ltd. (2005). *Glossary.* Retrieved March 28, 2006, from

http://www.oxagen.co.uk/pages/media_investors/6/

Pfizer. (2006). *Pfizer 2005 financial statement* (Financial Statement). Pfizer Inc.

Retrieved 10 March 2006,

Ramanathan, C. S., & Davison, D. B. (2002). Pharmaceutical bioinformatics and drug

discovery. In C. W. Sensen (Ed.), *Essentials of genomics and bioinformatics* (pp.

105-126). Weinheim: WILEY-VCH Verlag GmbH.

Researcher A. (2006). In Herron P. (Ed.), *Telephone interview*

Researcher B. (2006). In Herron P. (Ed.), *Telephone interview*

Rogers, E. M. (2003). *Diffusion of innovations* (5th , Free Press trade pbk. ed.). New

York: Free Press.

Roses, A. D., Burns, D. K., Chissoe, S., Middleton, L., & Jean, P. S. (2005). Disease-

specific target selection: A critical first step down the right road. *Drug Discovery

Today, 10*(3), 177-189.

Roses, A. D. (2000). Pharmacogenetics and the practice of medicine. *Nature, 405*(6788),

857-865.

Sachidanandam, R., Weissman, D., Schmidt, S. C., & et al. (2001). A map of human

genome sequence variation containing 1.42 million single nucleotide

polymorphisms. *Nature, 409*, 928-933.

Scherer, F. M. (2004). The pharmaceutical industry - prices and progress. *New England Journal of Medicine, 351*(9), 927-932.

Shannon, C. E., & Weaver, W. (1949). *The mathematical theory of communication.* Urbana and Chicago: University of Illinois Press.

Smith, I. (2004). *Historical notes about the cost of hard drive storage space.* Retrieved March 25, 2006, from http://www.littletechshoppe.com/ns1625/winchest.html

Smithies, O. (1955). Zone electrophoresis in starch gels: Group variations in the serum proteins of normal human adults. *Biochem. J., 61*, 629-641.

Swanson, D. R. (2006). Atrial fibrillation in athletes: Implicit literature-based connections suggest that overtraining and subsequent inflammation may be a contributory mechanism. *Medical Hypotheses, 66*(6), 1085-1092.

Swanson, D. R. (2001). *On the fragmentation of knowledge, the connection explosion, and assembling other people's ideas.* Retrieved September 28, 2006, from http://www.asis.org/Bulletin/Mar-01/swanson.html

Swanson, D. R. (1990). Somatomedin-C and arginine - implicit connections between mutually isolated literatures. *Perspectives in Biology and Medicine, 33*(2), 157-186.

Swanson, D. R. (1988a). Historical note - information-retrieval and the future of an illusion. *Journal of the American Society for Information Science, 39*(2), 92-98.

Swanson, D. R. (1988b). Migraine and magnesium - 11 neglected connections. *Perspectives in Biology and Medicine, 31*(4), 526-557.

Swanson, D. R. (1986a). Fish oil, raynauds syndrome, and undiscovered public

knowledge. *Perspectives in Biology and Medicine, 30*(1), 7-18.

Swanson, D. R. (1986b). Undiscovered public knowledge. *Library Quarterly, 56*(2), 103-

118.

Swanson, D. R. (1960). Searching natural language text by computer. *Science,*

132(3434), 1099-1104.

Theeramunkong, T. (2004). Applying passage in web text mining. *International Journal*

of Intelligent Systems, 19(1-2), 149-158.

US Food and Drug Administration Center for Drug Evaluation and Research. (2005a).

CDER new molecular entity (NME) drug and new biologic approvals in calendar

year 2004. Retrieved March 10, 2006, from

http://www.fda.gov.libproxy.lib.unc.edu/cder/rdmt/nmecy2004.htm

US Food and Drug Administration Center for Drug Evaluation and Research. (2005b).

CDER new molecular entity (NME) drug and new biologic approvals in calendar

year 2005. Retrieved March 10, 2006, from

http://www.fda.gov.libproxy.lib.unc.edu/cder/rdmt/nmecy2005.htm

Venkatesh, V., Morris, M. G., Davis, G. B., & Davis, F. D. (2003). User acceptance of

information technology: Toward a unified view. *MIS Quarterly, 27*(3), 425-478.

Weiss, S. M., Indurkhya, N., Zhang, T., & Damerau, F. J. (2004). *Text mining: Predictive*

methods for analyzing unstructured information. New York: Springer-Verlag.

Ziman, J. M. (1980). The proliferation of scientific literature - a natural process. *Science,*

208(4442), 369-371.

www.ingramcontent.com/pod-product-compliance
Lightning Source LLC
Chambersburg PA
CBHW081549220326
41598CB00036B/6619